AF765163

Powerplant

Test Guide

The "Fast-Track" to Study for and Pass the Aviation Maintenance Technician Powerplant Knowledge Test

2001

- **Effective June 19, 2000**
- **Organized in numerical order**
- **Questions, all FAA figures, Explanations, Answers and References arranged in the Fast-Track manner to help speed learning and retention**
- **Plus... an Oral & Practical Study Guide**

Aviation Supplies & Academics, Inc.
Newcastle, Washington

Powerplant Test Guide
2001 Edition

Aviation Supplies & Academics, Inc.
7005 132nd Place SE
Newcastle, WA 98059-3153
www.asa2fly.com

FAA questions herein are from United States government sources and contain current information as of: June 19, 2000.

Printed in the United States of America

01 00 5 4 3 2 1

For information, write or call:

ASA, Inc.
7005 132nd Place SE
Newcastle, WA 98059-3153
Voice: (425) 235-1500
Fax: (425) 235-0128
E-mail: asa@asa2fly.com

ASA-AMP-01

ISBN 1-56027-398-4

Contents

About the Author

Dale Crane, the author of the *Fast-Track Test Guides*, has been involved in aviation for more than 50 years as a mechanic, pilot, engineer, flight instructor, mechanic school instructor and director, mechanic examiner, and aviation writer.

He began his career in the U.S. Navy as a mechanic and flight engineer in PBYs. After World War II ended, he attended Parks Air College, majoring in Aviation Maintenance Engineering.

For 10 years after college, Mr. Crane worked at TEMCO Aircraft Corporation as an instrument overhaul mechanic, instrument shop manager, and flight test instrumentation engineer.

Following this, he spent 16 years as an instructor, then Director of the Aviation Maintenance Technician School of LeTourneau College.

For the past 20 years, he has been active as a writer of aviation technical materials and a consultant in developing aviation training programs.

He participated with the Federal Aviation Administration in the Aviation Mechanic Occupation Study (The David Allen Study) and the Aviation Mechanic Textbook Study.

Dale Crane holds the following FAA credentials:

- Airframe and Powerplant Mechanic
- Designated Mechanic Examiner
- Commercial Pilot
- Flight Instructor — Airplanes
- Advanced and Instrument Ground Instructor

For his 50 years of service in and contributions to the aviation maintenance industry, and the recognition of his peers for leadership excellence, Dale Crane has received the FAA's Charles Taylor "Master Mechanic" Award.

Other ASA Books by Dale Crane

Dictionary of Aeronautical Terms

Aviation Mechanic Handbook

Fast-Track Test Guides for Aviation Maintenance Technicians
- *General*
- *Airframe*
- *Powerplant*

Inspection Authorization Test Prep

Aviation Maintenance Technician Series:
- *General*
- *Airframe–Structures*
- *Airframe–Systems*
- *Powerplant*
- *Curriculum Guide*

Oral & Practical Exam Guide

The Fast-Track Method

The *Fast-Track* method of studying turns a multiple-choice examination into a study aid. It helps you learn the material in the shortest possible time, and you learn it in a way that you retain it.

The questions and the choices are duplicated directly from the FAA's website, and a clear explanation is given directly below the question.

To use the *Fast-Track* method, read the question, select your choice for the correct answer, then read the explanation without having to turn the page.

At the bottom of each page is the letter of the correct choice and the source upon which the FAA based the question (the Subject Matter Knowledge Code—*see* Page xv).

Free Test Updates for the One-Year Lifecycle of the Book

The FAA releases a new test database each spring, and makes amendments to this database approximately twice a year. ASA stays abreast of all changes to the tests, and posts these changes on the ASA website as a Test Update. ASA publishes Test Books each July. Visit the ASA website before taking your test to be certain you have all the current information: www.asa2fly.com

Preface

Employers of newly licensed aircraft mechanics have, for years, viewed the FAA A&P Knowledge Tests as a poor method for qualifying modern aircraft mechanics.

A&P schools have had the problem of teaching modern engines and systems, only to have to spend additional time prepping the student for the exam on the large radial engines and their components, early atmospheric control systems, and even repair procedures that a mechanic is not allowed to make.

The mechanic applicant is forced to memorize facts and figures about systems and components for which he has no frame of reference.

Hardly anyone likes the FAA Knowledge Tests, but we have them and will continue to have them.

In 1979, the FAA released "typical" examination questions as Advisory Circulars 65-20, 65-21, and 65-22 to help applicants prepare for the written exams. These have been superseded by the FAA written test books, and now, the FAA Knowledge Tests on computer.

I wrote the answers and explanations for these questions, and they did help applicants prepare for the FAA tests. But they had a serious drawback: the format in which they were presented proved to be awkward.

These first answer and explanation books had the questions in the front and the answers and explanations in the back. To compound the inefficiency, the letter for the correct choice was placed so prominently that it was the first thing seen. Choosing the correct letter (the right answer) does not encourage learning the explanation.

The *Fast-Track Test Guides* have corrected these problems and have proven to be the most effective way to study for an FAA A&P Knowledge Test.

The question and the choices are duplicated directly from the FAA Question Bank, and a clear explanation is given directly below the question.

Read the question, select your choice for the correct answer, then read the explanation without having to turn the page.

At the bottom of each page, where it is easy to see but where it does not encourage you to skip the explanation, is the letter of the correct choice for each question on that page, the subject matter knowledge code, and the source from which the FAA took the question.

The *Fast-Track* method of presenting the material allows you to learn the material in the shortest length of time, and it will help you retain the facts you learn.

The ASA *Fast-Track Test Guides* have included, as an important extra feature, typical Oral Questions and typical Practical Projects. These will give you an idea of the questions you will be asked and the projects you will be given to demonstrate your skills and reasoning.

ASA is dedicated to providing you with training materials that will help you become an A&P mechanic and to keep you up to date in this fascinating field. We welcome your criticism and suggestions so we can provide the materials you need.

Dale Crane

Test Code	Test Name	Number of Questions	Time (Hours) Allotted
AMP	Aviation Mechanic Powerplant	100	2.0

Instructions
Excerpt from FAA-G-8082-3

Introduction

What is required to become a skilled and effective airframe and powerplant (A&P) aviation mechanic? Although some individuals possess more knowledge and skills than others, no one is a natural-born aviation mechanic. Competent aviation mechanics become so through study, training, and experience.

This knowledge test guide will answer most of your questions about taking an aviation mechanic general, airframe, or powerplant knowledge test by covering the following areas: knowledge test eligibility requirements; knowledge areas on the tests; descriptions of the tests; process for taking a knowledge test; use of test aids and materials; cheating or other unauthorized conduct; validity of Airman Test Reports; and retesting procedures.

This guide will help in preparing you to take one or all of the following tests.

Aviation Mechanic—General

Aviation Mechanic—Airframe

Aviation Mechanic—Powerplant

This guide is not offered as an easy way to obtain the necessary information for passing the knowledge tests. Rather, the intent of this guide is to define and narrow the field of study to the required knowledge areas included in the tests.

Knowledge Test Eligibility Requirements

The general qualifications for an aviation mechanic certificate require you to have a combination of experience, knowledge, and skill. If you are pursuing an aviation mechanic certificate with airframe and powerplant ratings, you should review the appropriate sections of Title 14 of the Code of Federal Regulations (14 CFR) Part 65 for detailed information pertaining to eligibility requirements. Further information may be obtained from the nearest Flight Standards District Office (FSDO).

Before taking the certification knowledge and practical tests, you must meet the eligibility requirements. The determination of eligibility of applicants for the general, airframe, and powerplant tests is made on the basis of one of the following options:

1. **Civil and/or military experience**. (See 14 CFR Part 65, Subpart A—General, and Subpart D—Mechanics.) If you believe you are qualified to exercise this option, you must have your experience evaluated and certified by an FAA Aviation Safety Inspector (Airworthiness). If the inspector determines that you have the required experience, two FAA Forms 8610-2, Airman Certificate and/or Rating Application, are completed. These forms are issued, and MUST be presented along with appropriate identification to take the corresponding knowledge tests. Your eligibility to test does not expire.
2. **Graduation from an FAA-certificated Aviation Maintenance Technician School (AMTS)**. Depending upon the testing facility affiliation[1], a graduation certificate, certificate of completion, or an FAA Form 8610-2, Airman Certificate and/or Rating Application (properly endorsed) is required, along with proper identification.

If you are taking the tests at a computer testing center and the practical testing is administered by a designated mechanic examiner (DME), and BOTH are affiliated with the AMTS, a copy of the graduation certificate or certificate of completion (along with proper identification) may be all that you are required to present. In this case, the school, the testing center, the DME, and the local FSDO will all be involved and know what authorization is needed. On the other hand, if either one, or both the testing center and the DME are NOT affiliated with the AMTS, then FAA Form 8610-2 is required.

[1] Affiliation is a procedural arrangement to provide for graduates to take the knowledge and practical tests. The arrangement requirements are agreed to by a particular school, testing center, and designated mechanic examiner (DME), having also been approved by the supervising FAA FSDO.

Knowledge Areas on the Tests

Aviation mechanic tests are comprehensive because they must test your knowledge in many subject areas. The subject areas for the tests are the same as the required AMTS curriculum subjects listed in 14 CFR Part 147, Appendixes B, C, and D. However, the subject area titled "Unducted Fans" (in Appendix D) is not a tested subject at this time. The terms used in 14 CFR Part 147, Appendixes B, C, and D are defined in 14 CFR Part 147, Appendix A.

Description of the Tests

All test questions are the objective, multiple-choice type. Each question can be answered by the selection of a single response. Each test question is independent of other questions; therefore, a correct response to one does not depend upon, or influence, the correct response to another.

The aviation mechanic general test contains 50 questions, and you are allowed 1.5 hours to complete the test.

The aviation mechanic airframe and aviation mechanic powerplant tests contain 100 questions, and you are allowed 2 hours to complete each test.

Communication between individuals through the use of words is a complicated process. In addition to being an exercise in the application and use of aeronautical knowledge, a knowledge test is also an exercise in communication since it involves the use of the written language. Since the tests involve written rather than spoken words, communication between the test writer and the person being tested may become a difficult matter if care is not exercised by both parties. Consequently, considerable effort is expended to write each question in a clear, precise manner. Make sure you read the instructions given with the test, as well as the statements in each test item.

When taking a test, keep the following points in mind:

1. Answer each question in accordance with the latest regulations and guidance publications.
2. Read each question carefully before looking at the possible answers. You should clearly understand the problem before attempting to solve it.
3. After formulating an answer, determine which choice corresponds with that answer. The answer chosen should completely resolve the problem.
4. From the answers given, it may appear that there is more than one possible answer; however, there is only one answer that is correct and complete. The other answers are either incomplete, erroneous, or represent common misconceptions.
5. If a certain question is difficult for you, it is best to mark it for review and proceed to the next question. After you answer the less difficult questions, return to those which you marked for review and answer them. The review marking procedure will be explained to you prior to starting the test. Although the computer should alert you to unanswered questions, make sure every question has an answer recorded. This procedure will enable you to use the available time to maximum advantage.
6. When solving a calculation problem, select the answer closest to your solution. The problem has been checked several times by various individuals; therefore, if you have solved it correctly, your answer will be closer to the correct answer than any of the other choices.

Process for Taking a Knowledge Test

The Federal Aviation Administration (FAA) has available hundreds of computer testing centers worldwide. These testing centers offer the full range of airman knowledge tests including recreational through airline transport pilot, parachute rigger, mechanic, and mechanic examiner tests. Refer to the list of computer testing designees (CTDs) at the end of this section.

The first step in taking a knowledge test is the registration process. You may either call the testing centers' 1-800 numbers or simply take the test on a walk-in basis. If you choose to use the 1-800 number to register, you will need to select a testing center, schedule a test date, and make financial arrangements for test payment. You may register for tests several weeks in advance, and you may cancel your appointment according to the CTD's cancellation policy. If you do not follow the CTD's cancellation policies, you could be subject to a cancellation fee.

The next step in taking a knowledge test is providing proper identification. You should determine what knowledge test prerequisites are necessary before going to the computer testing center. Your instructor or local FSDO can assist you with what documentation to take to the testing facility. Testing center personnel will not begin the test until your identification is verified. A limited number of tests do not require authorization.

Acceptable forms of authorization are:

- FAA Form 8610-2.
- A graduation certificate or certificate of completion to an affiliated testing center as previously explained.
- An original (not photocopy) failed Airman Test Report, passing Airman Test Report, or expired Airman Test Report.

Before you take the actual test, you will have the option to take a sample test. The actual test is time limited; however, you should have sufficient time to complete and review your test.

Upon completion of the knowledge test, you will receive your Airman Test Report, with the testing center's embossed seal, which reflects your score.

The Airman Test Report lists the subject matter knowledge codes for questions answered incorrectly. The total number of subject matter knowledge codes shown on the Airman Test Report is not necessarily an indication of the total number of questions answered incorrectly. The subject matter knowledge codes that refer to the knowledge areas are listed in the next section of this book. Study these knowledge areas to improve your understanding of the subject matter.

The Airman Test Report must be presented to the examiner prior to taking the practical test. During the oral portion of the practical test, the examiner is required to evaluate the noted areas of deficiency.

Should you require a duplicate Airman Test Report due to loss or destruction of the original, send a signed request accompanied by a check or money order for $1 payable to the FAA. Your request should be sent to the Federal Aviation Administration, Airmen Certification Branch, AFS-760, P.O. Box 25082, Oklahoma City, OK 73125.

Use of Test Aids and Materials

Airman knowledge tests require applicants to analyze the relationship between variables needed to solve aviation problems, in addition to testing for accuracy of a mathematical calculation. The intent is that all applicants are tested on concepts rather than rote calculation ability. It is permissible to use certain calculating devices when taking airman knowledge tests, provided they are used within the following guidelines. The term "calculating devices" is interchangeable with such items as calculators, computers, or any similar devices designed for aviation-related activities.

1. Guidelines for use of test aids and materials. The applicant may use test aids and materials within the guidelines listed below, if actual test questions or answers are not revealed.
    a. Applicants may use test aids, such as a calculating device that is directly related to the test. In addition, applicants may use any test materials provided with the test.
    b. The test proctor may provide a calculating device to applicants and deny them use of their personal calculating device if the applicant's device does not have a screen that indicates all memory has been erased. The test proctor must be able to determine the calculating device's erasure capability. The use of calculating devices incorporating permanent or continuous type memory circuits without erasure capability is prohibited.
    c. The use of magnetic cards, magnetic tapes, modules, computer chips, or any other device upon which prewritten programs or information related to the test can be stored and retrieved is prohibited. Printouts of data will be surrendered at the completion of the test if the calculating device used incorporates this design feature.
    d. The use of any booklet or manual containing instructions related to the use of the applicant's calculating device is not permitted.
    e. Dictionaries are not allowed in the testing area.
    f. The test proctor makes the final determination relating to test materials and personal possessions that the applicant may take into the testing area.

Continued

2. Guidelines for dyslexic applicant's use of test aids and materials. A dyslexic applicant may request approval from the local Flight Standards District Office (FSDO) to take an airman knowledge test using one of the three options listed in preferential order:
   a. Option One. Use current testing facilities and procedures whenever possible.
   b. Option Two. Applicants may use a Franklin Speaking Wordmaster® to facilitate the testing process. The Wordmaster® is a self-contained electronic thesaurus that audibly pronounces typed in words and presents them on a display screen. It has a built-in headphone jack for private listening. The headphone feature will be used during testing to avoid disturbing others.
   c. Option Three. Applicants who do not choose to use the first or second option may request a test proctor to assist in reading specific words or terms from the test questions and supplement material. In the interest of preventing compromise of the testing process, the test proctor should be someone who is non-aviation oriented. The test proctor will provide reading assistance only, with no explanation of words or terms. The Airman Testing Standards Branch, AFS-630, will assist in the selection of a test site and test proctor.

Cheating or Other Unauthorized Conduct

Computer testing centers are required to follow strict security procedures to avoid test compromise. These procedures are established by the FAA and are covered in FAA Order 8080.6, Conduct of Airman Knowledge Tests. The FAA has directed testing centers to terminate a test at any time a test proctor suspects a cheating incident has occurred. An FAA investigation will then be conducted. If the investigation determines that cheating or other unauthorized conduct has occurred, then any airman certificate or rating that you hold may be revoked, and you will be prohibited for 1 year from applying for or taking any test for a certificate or rating under 14 CFR Part 65.

Validity of Airman Test Reports

Airman Test Reports are valid for the 24-calendar month period preceding the month you complete the practical test. If the Airman Test Report expires before completion of the practical test, you must retake the knowledge test.

Retesting Procedures

If you receive a grade lower than a 70 percent and wish to retest, you must present the following to testing center personnel:

- failed Airman Test Report; or
- if you apply within 30 days, a failed Airman Test Report with an endorsement from a mechanic certificate holder with the same rating(s) you are testing for, certifying that additional instruction has been given, and that you have been found competent to pass the test.

If you decide to retake the test in anticipation of a better score, you may retake the test after 30 days from the date your last test was taken. The FAA will not allow you to retake a passed test before the 30-day period has lapsed. Prior to retesting, you must give your current Airman Test Report to the test proctor. The last test taken will reflect the official score.

Computer Testing Designees

The following is a list of the computer testing designees authorized to give FAA knowledge tests. This list should be helpful in case you choose to register for a test or simply want more information.

Computer Assisted Testing Service (CATS)

1849 Old Bayshore Highway
Burlingame, CA 94010
Applicant inquiry and test registration: 1-800-947-4228
From outside the U.S.: (650) 259-8550

LaserGrade Computer Testing

16209 S.E. McGillivray, Suite L
Vancouver, WA 98683
Applicant inquiry and test registration: 1-800-211-2753 or 1-800-211-2754
From outside the U.S.: (360) 896-9111

The latest listing of computer testing center locations may be obtained through the FAA website: http://afs600.faa.gov, then select AFS630, Airman Certification, Computer Testing Sites.

Subject Matter Knowledge Codes

To determine the knowledge area in which a particular question was incorrectly answered, compare the subject matter code(s) on the Knowledge Test Report, to the subject matter outline that follows. The total number of test items missed may differ from the number of subject matter codes shown on the test report, since you may have missed more than one question in a certain subject matter code. This list of Subject Matter Knowledge Codes is taken from AC 60-25D.

Reciprocating Engines—AC 65-9A, AC 65-12A, 14 CFR part 43, AP, JSPT, AMT-P

A01 Inspect and repair a radial engine
A02 Overhaul reciprocating engine
A03 Inspect, check, service, and repair reciprocating engines and engine installations
A04 Install, troubleshoot, and remove reciprocating engines

Turbine Engines—AC 65-9A, AC 65-12A, AC 65-15A, 14 CFR part 33, AP, AGTP, JSPT

B01 Overhaul turbine engine
B02 Inspect, check, service, and repair turbine engines and turbine engine installations
B03 Install, troubleshoot, and remove turbine engines

Engine Inspection—AC 65-9A, AC 65-12A, AC 39-7B, AC 43.13-1B, 14 CFR part 23, 14 CFR part 33, 14 CFR part 43, 14 CFR part 65, ABS, AP, JSGT, JSPT

C01 Perform powerplant conformity and airworthiness inspections
DXX Reserved
EXX Reserved
FXX Reserved
GXX Reserved

Engine Instrument Systems—AC 65-12A, AC 65-15A, AC 20-88A, 14 CFR part 65, AMR, AP, AGTP, JSPT, ATD

H01 Troubleshoot, service, and repair electrical and mechanical fluid rate-of-flow indicating systems
H02 Inspect, check, service, troubleshoot, and repair electrical and mechanical engine temperature, pressure, and RPM indicating systems

Engine Fire Protection Systems—AC 65-9A, AC 65-12A, ABS, AMR, AP, JSPT

I01 Inspect, check, service, troubleshoot, and repair engine fire detection and extinguishing systems

Engine Electrical Systems—AC 65-9A, AC 65-12A, AC 65-15A, AC 43.13-1B, 14 CFR part 23, 14 CFR part 25, AEE, AP, JSGT, JSPT

J01 Repair engine electrical system components
J02 Install, check, and service engine electrical wiring, controls, switches, indicators, and protective devices

Lubrication Systems—AC 65-12A, AC 65-15A, 14 CFR part 33, AP, AGTP, JSPT, AMT-P

K01 Identify and select lubricants
K02 Repair engine lubrication system components
K03 Inspect, check, service, troubleshoot, and repair engine lubrication systems

Continued

Ignition and Starting Systems—AC 65-12A, AC 65-15A, AEE, AP, AGTP, JSPT

L01 Overhaul magneto and ignition harness
L02 Inspect, service, troubleshoot, and repair reciprocating and turbine engine ignition systems and components
L03 Inspect, service, troubleshoot, and repair turbine engine electrical starting systems
L04 Inspect, service, and troubleshoot turbine engine pneumatic starting systems

Fuel Metering Systems—AC 65-9A, AC 65-12A, AP, AGTP, JSPT

M01 Troubleshoot and adjust turbine engine fuel metering systems and electronic engine fuel controls
M02 Overhaul carburetor
M03 Repair engine fuel metering system components
M04 Inspect, check, service, troubleshoot, and repair reciprocating and turbine engine fuel metering systems

Engine Fuel Systems—AC 65-9A, AC 65-12A, AC 43.13-1B, 14 CFR part 23, AP, JSPT

N01 Repair engine fuel system components
N02 Inspect, check, service, troubleshoot, and repair engine fuel systems

Induction and Engine Airflow Systems—AC 65-9A, AC 65-12A, AC 43.13-1B, AP, AGTP, JSPT, AMT-P

O01 Inspect, check, troubleshoot, service, and repair engine ice and rain control systems
O02 Inspect, check, service, troubleshoot, and repair heat exchangers, superchargers, and turbine engine airflow and temperature control systems
O03 Inspect, check, service, and repair carburetor air intake and induction manifolds

Engine Cooling Systems—AC 65-12A, ABS, AP, JSPT, AMT-P

P01 Repair engine cooling system components
P02 Inspect, check, troubleshoot, service, and repair engine cooling systems

Engine Exhaust and Reverser Systems—AC 65-12A, AC 43.13-1B, JSPT

Q01 Repair engine exhaust system components
Q02 Inspect, check, troubleshoot, service, and repair engine exhaust systems
Q03 Troubleshoot and repair engine thrust reverser systems and related components

Propellers—AC 65-9A, AC 65-12A, AC 43.13-1B, 14 CFR part 43, 14 CFR part 65, AP, ATD, APC, JSPT, AMT-P

R01 Inspect, check, service, and repair propeller synchronizing and ice control systems
R02 Identify and select propeller lubricants
R03 Balance propellers
R04 Repair propeller control system components
R05 Inspect, check, service, and repair fixed pitch, constant speed and feathering propellers, and propeller governing systems
R06 Install, troubleshoot, and remove propellers
R07 Repair aluminum alloy propeller blades

Auxiliary Power Units—DAT, TCAS, ATD, AGTP

T01 Inspect, check, service, and troubleshoot turbine-driven auxiliary power units

Abbreviations and References

The following abbreviations are used to identify the reference associated with each test question.

ABS	Aircraft Basic Science—Glencoe Division, Macmillan/McGraw-Hill Publication Company
AC	Advisory Circular
AEE	Aircraft Electricity and Electronics —Glencoe Division, Macmillan/McGraw-Hill Publication Company
AMR	Aircraft Maintenance and Repair—Glencoe Division, Macmillan/McGraw-Hill Publication Company
AMT-G	Aviation Maintenance Technician Series General—Aviation Supplies & Academics (ASA), Inc.
AMT-P	Aviation Maintenance Technician Series Powerplant—Aviation Supplies & Academics (ASA), Inc.
AP	Aircraft Powerplants—Glencoe Division, Macmillan/McGraw-Hill Publication Company
DAT	Dictionary of Aeronautical Terms—Aviation Supplies & Academics (ASA), Inc.
TCAS	Transport Category Aircraft Systems—Jeppesen Sanderson, Inc.
APC	Aircraft Propellers and Controls—Jeppesen Sanderson, Inc.
ATD	Aircraft Technical Dictionary—Jeppesen Sanderson, Inc.
JSGT	A & P Technician General Textbook—Jeppesen Sanderson, Inc.
JSPT	A & P Technician Powerplant Textbook—Jeppesen Sanderson, Inc.
AGTP	Aircraft Gas Turbine Powerplants—Jeppesen Sanderson, Inc.
14 CFR	Title 14 of the Code of Federal Regulations (part or § [section]) — Government Printing Office (GPO)
PSG	A & P Technician Powerplant Study Guide—Jeppesen Sanderson, Inc.

Powerplant Test Questions, Explanations, Answers & References

At the bottom of every page are the answers with coded items as explained below:

Code	Explanation
8001	This is the number which corresponds to the question number in the Question section of this Test Guide and the FAA Test.
[ ]	The brackets enclose the letter answer selected by ASA's researchers.
( )	The parentheses enclose the appropriate Subject Matter Knowledge Code.
DAT, AC, etc.	The reference following the Subject Matter Knowledge Code is the source from which the answer was derived. The meanings of these abbreviations are found on Page xvii.

8001. Which statement is true regarding bearings used in high-powered reciprocating aircraft engines?

A—The outer race of a single-row, self-aligning ball bearing will always have a radius equal to the radius of the balls.
B—There is less rolling friction when ball bearings are used than when roller bearings are employed.
C—Crankshaft bearings are generally of the ball-type due to their ability to withstand extreme loads without overheating.

The smaller contact area of a ball bearing causes it to produce less rolling friction than a roller bearing.

Ball bearings are used in high-powered reciprocating engines, where keeping friction to a minimum is important.

Ball bearings can be designed and installed in such a way that they reduce friction in axial loads as well as in radial loads.

8002. A condition that can occur in radial engines but is unlikely to occur in horizontally opposed engines is

A—zero valve clearance.
B—valve overlap.
C—hydraulic lock.

Radial and inverted engines have some cylinders below the crankcase, and when the engine is idle, oil will leak from the crankcase, past the piston rings, and fill the combustion chamber. This condition is called a hydraulic lock.

If this oil is not removed before the engine is started, the piston will move against the noncompressible oil and cause serious damage.

8003. Which condition would be the least likely to be caused by failed or failing engine bearings?

A—Excessive oil consumption.
B—High oil temperatures.
C—Low oil temperatures.

All of the alternatives except low oil temperature would likely be caused by failed or failing engine bearings in a reciprocating engine.

Low oil temperature would be the least likely of these alternatives.

8004. What is the principal advantage of using propeller reduction gears?

A—To enable the propeller RPM to be increased without an accompanying increase in engine RPM.
B—To enable the engine RPM to be increased with an accompanying increase in power and allow the propeller to remain at a lower, more efficient RPM.
C—To enable the engine RPM to be increased with an accompanying increase in propeller RPM.

The horsepower produced by a reciprocating engine is determined by its RPM. The higher the RPM, the greater the power. But the efficiency of a propeller decreases as the blade tip speed approaches the speed of sound.

In order to get the best of both conditions, many of the more powerful aircraft engines drive the propeller through a set of reduction gears.

Reduction gears allow the engine to turn fast enough to develop the required power. At the same time, the propeller tip speed is kept low enough that the tips do not approach the speed of sound.

8005. Which of the following will decrease volumetric efficiency in a reciprocating engine?

1. Full throttle operation.
2. Low cylinder head temperatures.
3. Improper valve timing.
4. Sharp bends in the induction system.
5. High carburetor air temperatures.

A—2, 4, and 5.
B—1, 2, 3, and 4.
C—3, 4, and 5.

The volumetric efficiency of a reciprocating engine is the ratio of the weight of the fuel-air charge taken into the cylinder, to the weight of a charge that would completely fill the entire volume of the cylinder at the same pressure.

Anything that decreases the weight of the air entering the cylinder decreases the volumetric efficiency. Improper valve timing, sharp bends in the induction system, and high carburetor air temperature will all decrease the volumetric efficiency.

8006. Which of the following is a characteristic of a thrust bearing used in most radial engines?

A—Tapered roller.
B—Double-row ball.
C—Deep-groove ball.

Deep-groove ball bearings are used as the thrust bearing in most radial engines. This type of bearing is the best of those listed for reducing friction while carrying both thrust and radial loads.

Answers

8001 [B] (A01) AMT-P	8002 [C] (A01) AMT-P	8003 [C] (A01) AMT-P	8004 [B] (A01) AMT-P
8005 [C] (A01) AMT-P	8006 [C] (A01) AMT-P		

8007. Which bearing is least likely to be a roller or ball bearing?

A—Rocker arm bearing (overhead valve engine).
B—Master rod bearing (radial engine).
C—Crankshaft main bearing (radial engine).

The master rod bearing in a radial engine is always a plain bearing.

Rocker arm bearings may be either ball, roller, or plain type and the crankshaft main bearings for radial engines are usually ball bearings.

8008. The operating temperature valve clearance of a radial engine as compared to cold valve clearance is

A—greater.
B—less.
C—the same.

When a radial engine is operating, the cast aluminum alloy cylinder head expands far more than the steel push rod. As the cylinder head expands, the rocker arm moves away from the cam ring and the hot, or running, valve clearance becomes much greater than the cold clearance.

8009. A nine-cylinder engine with a bore of 5.5 inches and a stroke of 6 inches will have a total piston displacement of

A—740 cubic inches.
B—1,425 cubic inches.
C—1,283 cubic inches.

The piston displacement of a reciprocating engine is the total volume swept by the pistons in one revolution of the crankshaft.

Find the piston displacement of one cylinder by multiplying the area of the piston in square inches by the stroke, which is measured in inches.

The total piston displacement is the volume of one cylinder, measured in cubic inches, multiplied by the number of cylinders.

$Area = 0.7854 \times bore^2$
$= 0.7854 \times 30.25$
$= 23.75$ *square inches*

$Volume = piston\ area \times stroke$
$= 23.75 \times 6$
$= 142.55$ *cubic inches*

$Piston\ displacement = volume \times number\ of\ cylinders$
$= 142.55 \times 9$
$= 1{,}282.95$ *cubic inches*

8010. The five events of a four-stroke cycle engine in the order of their occurrence are

A—intake, ignition, compression, power, exhaust.
B—intake, power, compression, ignition, exhaust.
C—intake, compression, ignition, power, exhaust.

The five events that take place in a reciprocating engine during each cycle of its operation are:

Intake—*The fuel-air mixture is taken into the cylinder.*
Compression—*The fuel-air mixture is compressed as the piston moves upward (outward) in the cylinder.*
Ignition—*As the piston nears the top of its stroke, an electrical spark ignites the mixture so it burns and releases its energy.*
Power—*As the fuel-air mixture burns, it forces the piston downward. This movement of the piston rotates the crankshaft and performs useful work.*
Exhaust—*After the piston has reached the bottom of its stroke and done the most of its useful work, the piston pushes upward, forcing the burned gases out of the cylinder.*

8011. The primary concern in establishing the firing order for an opposed engine is to

A—provide for balance and eliminate vibration to the greatest extent possible.
B—keep power impulses on adjacent cylinders as far apart as possible in order to obtain the greatest mechanical efficiency.
C—keep the power impulses on adjacent cylinders as close as possible in order to obtain the greatest mechanical efficiency.

The firing order of an opposed engine is designed to provide for balance and to eliminate vibration as much as possible.

8012. If fuel/air ratio is proper and ignition timing is correct, the combustion process should be completed

A—20 to 30° before top center at the end of the compression stroke.
B—when the exhaust valve opens at the end of the power stroke.
C—just after top center at the beginning of the power stroke.

The ignition of the fuel-air mixture in the cylinder of a reciprocating engine is timed so it occurs when the piston is about 20 to 30 degrees of crankshaft rotation before reaching top center on the compression stroke.

If the mixture ratio and ignition timing are both correct, the fuel-air mixture will be all burned shortly after the

Answers

8007 [B] (A01) AMT-P	8008 [A] (A01) AMT-P	8009 [C] (A01) AMT-P	8010 [C] (A01) AMT-P
8011 [A] (A02) AMT-P	8012 [C] (A02) AMT-P		

piston passes over top center. The expanding gases caused by absorbing heat from the burning mixture will exert the maximum amount of push on the descending piston during the power stroke.

8013. Grinding the valves of a reciprocating engine to a feather edge is likely to result in

A—normal operation and long life.
B—excessive valve clearance.
C—preignition and burned valves.

If a valve is ground with a feather edge (a thin edge) the heat in the cylinder will cause the thin area to glow red hot and this will ignite the fuel-air mixture before the correct time for ignition. This will result in preignition and burned valves.

8014. Which statement is correct regarding engine crankshafts?

A—Moveable counterweights serve to reduce the dynamic vibrations in an aircraft reciprocating engine.
B—Moveable counterweights serve to reduce the torsional vibrations in an aircraft reciprocating engine.
C—Moveable counterweights are designed to resonate at the natural frequency of the crankshaft.

Torsional vibration caused by firing impulses of the engine are minimized by the installation of moveable counterweights suspended from certain crank cheeks. These moveable counterweights, called dynamic dampers, rock back and forth and act as pendulums, changing the resonant frequency of the rotating elements, thus reducing the torsional vibration.

8015. On which strokes are both valves on a four-stroke cycle reciprocating aircraft engine open?

A—Power and exhaust.
B—Intake and compression.
C—Exhaust and intake.

Both the intake and exhaust valve are open at the same time only during the period of valve overlap.

Valve overlap occurs at the end of the exhaust stroke and the beginning of the intake stroke. The intake valve opens a few degrees of crankshaft rotation before the piston reaches the top of the exhaust stroke. The exhaust valve remains open until the piston has moved down a few degrees of crankshaft rotation on the intake stroke.

8016. Master rod bearings are generally what type?

A—Plain.
B—Roller.
C—Ball.

Master rods used in radial engines have plain bearings in both their big end that fits around the throw of the crankshaft and the small end that fits around the wrist pin in the piston.

8017. The actual power delivered to the propeller of an aircraft engine is called

A—friction horsepower.
B—brake horsepower.
C—indicated horsepower.

The actual horsepower delivered to the propeller of an aircraft engine is called brake horsepower. This name is used because brake horsepower was originally measured with a prony brake loading the engine with mechanical friction.

Modern measurements of brake horsepower are made with a dynamometer which loads the engine with electrical or fluid-flow opposition.

8018. Cam-ground pistons are installed in some aircraft engines to

A—provide a better fit at operating temperatures.
B—act as a compensating feature so that a compensated magneto is not required.
C—equalize the wear on all pistons.

A cam-ground piston is one whose diameter is a few thousandths of an inch greater in a plane perpendicular to the wrist pin boss than it is parallel to the boss.

When the piston reaches its operating temperature, the large mass of metal in the piston pin boss expands enough that the piston becomes round.

Since the piston is round at its operating temperature, it provides a better seal than it would if it were round while cold and expanded to an out-of-round condition when hot.

8019. Using the following information, determine how many degrees the crankshaft will rotate with both the intake and exhaust valves seated.

Intake opens 15° BTDC.
Exhaust opens 70° BBDC.
Intake closes 45° ABDC.
Exhaust closes 10° ATDC.

A—290°.
B—245°.
C—25°.

The intake valve closes 45° of crankshaft rotation after the piston passes bottom dead center, moving upward on the compression stroke.

Continued

Answers

8013 [C] (A02) AMT-P	8014 [B] (A02) AMT-P	8015 [C] (A02) AMT-P	8016 [A] (A02) AMT-P
8017 [B] (A02) AMT-P	8018 [A] (A02) AMT-P	8019 [B] (A02) AMT-P	

Both valves are closed at this point, and they both remain closed until the piston passes over top center and comes down to 70° before bottom dead center on the power stroke. At this time the exhaust valve opens.

Both valves are on their seats for 45° + 180° + 20°, or 245°.

8020. Some aircraft engine manufacturers equip their product with choked or taper-ground cylinders in order to

A—provide a straight cylinder bore at operating temperatures.
B—flex the rings slightly during operation and reduce the possibility of the rings sticking in the grooves.
C—increase the compression pressure for starting purposes.

Some aircraft engine cylinders are ground with the diameter at the top of the barrel, where it screws into the head, slightly smaller than the diameter in the center of the barrel. This is called choke grinding.

The large mass of the cylinder head expands more when heated than the smaller mass of the cylinder barrel, so the diameter of a choke-ground cylinder becomes uniform when the engine is at its operating temperature.

8021. An aircraft reciprocating engine using hydraulic valve lifters is observed to have no clearance in its valve-operating mechanism after the minimum inlet oil and cylinder head temperatures for takeoff have been reached. When can this condition be expected?

A—During normal operation.
B—When the lifters become deflated.
C—As a result of carbon and sludge becoming trapped in the lifter and restricting its motion.

There is no clearance in the valve-operating mechanism when an engine equipped with hydraulic valve lifters is operating normally and the minimum oil and cylinder-head temperatures for takeoff have been reached.

Hydraulic valve lifters are used because they remove all of the clearance between the rocker arm and the tip of the valve stem.

By keeping all of this clearance removed, the valves operate with less noise and less wear.

8022. What tool is generally used to measure the crankshaft rotation in degrees?

A—Dial indicator.
B—Timing disk.
C—Prop Protractor.

A top dead center indicator is used to show when the piston in cylinder number one is on top dead center.

A timing disk is clamped to the propeller shaft and positioned so the pointer, which is held straight up by a weight on one end, points to zero degrees.

As the crankshaft is rotated, the pointer indicates on the scale of the timing disk the number of degrees the crankshaft has rotated.

8023. If an engine with a stroke of 6 inches is operated at 2,000 RPM, the piston movement within the cylinder will be

A—at maximum velocity around TDC.
B—constant during the entire 360° of crankshaft travel.
C—at maximum velocity 90° after TDC.

The piston in a reciprocating engine is not moving when it is at the top and bottom of its stroke.

As it leaves top dead center, it accelerates from zero velocity to a maximum velocity, which is reached when it is 90° beyond top dead center. It then decelerates to zero velocity at bottom dead center.

8024. If the intake valve is opened too early in the cycle of operation of a four-stroke cycle engine, it may result in

A—improper scavenging of exhaust gases.
B—engine kickback.
C—backfiring into the induction system.

The intake valve opens when the piston is moving upward at the end of the exhaust stroke. Opening at this point allows the low pressure caused by the inertia of the exiting exhaust gases to assist in starting the fuel-air mixture flowing into the cylinder.

If the intake valve opens too early, some of the burning exhaust gases could flow into the intake manifold and ignite the mixture. This would cause a backfire in the induction system.

8025. Some cylinder barrels are hardened by

A—nitriding.
B—shot peening.
C—tempering.

The walls of an aircraft-engine cylinder are subjected to a great deal of wear as the iron piston rings rub against them.

The walls of some cylinders are treated to increase their hardness and resistance to wear. There are two methods of hardening these surfaces: hard-chrome-plating and nitriding.

Nitriding is a process in which the surface of the steel cylinder wall is changed into a hard nitride by an infusion of nitrogen from the ammonia gas used in the nitriding heat treatment process.

Answers

8020 [A] (A02) AMT-P	8021 [A] (A02) AMT-P	8022 [B] (A02) AMT-P	8023 [C] (A02) AMT-P
8024 [C] (A02) AMT-P	8025 [A] (A02) AMT-P		

8026. Which statement is correct regarding a four-stroke cycle aircraft engine?

A—The intake valve closes on the compression stroke.
B—The exhaust valve opens on the exhaust stroke.
C—The intake valve closes on the intake stroke.

The intake valve in a four-stroke-cycle aircraft engine closes somewhere around 60° after bottom center on the compression stroke.

The exhaust valve opens about 70° before bottom center on the power stroke.

The intake valve opens about 20° before top center on the exhaust stroke.

The exhaust valve closes about 15° after top center on the intake stroke.

8027. On which part of the cylinder walls of a normally operating engine will the greatest amount of wear occur?

A—Near the center of the cylinder where piston velocity is greatest.
B—Near the top of the cylinder.
C—Wear is normally evenly distributed.

In normal operation, an aircraft engine cylinder wears more at the top than in the center or at the bottom. This greater wear is caused by the heat of combustion decreasing the efficiency of the lubrication at the top of the cylinder.

8028. During overhaul, reciprocating engine exhaust valves are checked for stretch

A—with a suitable inside spring caliper.
B—with a contour or radius gauge.
C—by placing the valve on a surface plate and measuring its length with a vernier height gauge.

One recommended way of checking exhaust valves for stretch is by measuring the diameter of the valve stem with a vernier outside micrometer caliper at a point specified by the engine manufacturer. If the valve has stretched, the stem diameter will be smaller than it should be.

Another way of determining if a valve has been stretched is by using a valve radius gauge to see if the radius between the valve stem and head is the same radius the valve had when it was manufactured.

8029. When is the fuel/air mixture ignited in a conventional reciprocating engine?

A—When the piston has reached top dead center of the intake stroke.
B—Shortly before the piston reaches the top of the compression stroke.
C—When the piston reaches top dead center on the compression stroke.

Ignition occurs in a reciprocating engine somewhere around 30° of crankshaft rotation before the piston reaches top center on the compression stroke.

By timing the ignition to occur when the piston is in this position, the maximum pressure inside the cylinder is reached just after the piston passes over top center and starts down on the power stroke.

8030. Ignition occurs at 28° BTDC on a certain four-stroke cycle engine, and the intake valve opens at 15° BTDC. How many degrees of crankshaft travel after ignition does the intake valve open? (Consider one cylinder only.)

A—707°.
B—373°.
C—347°.

The crankshaft rotates 28° on the compression stroke after the ignition occurs.

The crankshaft rotates 180° on the power stroke.

The crankshaft rotates 165° on the exhaust stroke before the intake valve opens.

The total crankshaft rotation between the time ignition occurs and the time the intake valve opens is:

$28° + 180° + 165° = 373°.$

8031. What is the purpose of the safety circlet installed on some valve stems?

A—To hold the valve guide in position.
B—To hold the valve spring retaining washer in position.
C—To prevent valves from falling into the combustion chamber.

Some aircraft-engine poppet valves have a groove cut in their stem that is fitted with a safety circlet, a small snap ring that grips the valve stem in this groove.

If the tip of the valve stem should ever break off in operation, this safety circlet will contact the top of the valve guide and prevent the valve from dropping into the cylinder.

Answers

8026 [A] (A02) AMT-P	8027 [B] (A02) AMT-P	8028 [B] (A02) AMT-P	8029 [B] (A02) AMT-P
8030 [B] (A02) AMT-P	8031 [C] (A02) AMT-P		

8032. Valve overlap is defined as the number of degrees of crankshaft travel

A—during which both valves are off their seats.
B—between the closing of the intake valve and the opening of the exhaust valve.
C—during which both valves are on their seats.

Valve overlap is the number of degrees of crankshaft rotation that both the intake and exhaust valves are off their seat at the end of the exhaust stroke and the beginning of the intake stroke.

Valve overlap allows a greater charge of fuel-air mixture to be inducted into the cylinder.

8033. The valve clearance of an engine using hydraulic lifters, when the lifters are completely flat, or empty, should not exceed

A—0.00 inch.
B—a specified amount above zero.
C—a specified amount below zero.

Hydraulic valve lifters are used to keep all of the clearance out of the valve system when the engine is operating and the lifters are pumped up.

When the lifters are completely flat, there will be clearance in the system of a specified amount above zero.

8034. If the exhaust valve of a four-stroke cycle engine is closed and the intake valve is just closed, the piston is on the

A—intake stroke.
B—power stroke.
C—compression stroke.

The intake valve closes when the piston is moving upward on the compression stroke. At this time, the exhaust valve is already closed.

8035. How many of the following are factors in establishing the maximum compression ratio limitations of an aircraft engine?

1. Detonation characteristics of the fuel used.
2. Design limitations of the engine.
3. Degree of supercharging.
4. Spark plug reach.

A—Four.
B—Two.
C—Three.

The maximum compression ratio of an engine is limited by the ability of the engine to withstand detonation in its cylinders.

Of the alternatives given with this question, three of them are factors affecting the engine's ability to withstand detonation.

The detonation characteristics of the fuel used is a limiting factor. Fuels having a low critical pressure and temperature must not be used with high compression engines.

The design limitations of the engine are important, because engines that are not designed strong enough to withstand high cylinder pressures, must not have a high compression ratio.

The degree of supercharging is extremely important, because the cylinder pressures are a function of both the initial pressure in the cylinder (the pressure caused by the supercharger) and the compression ratio.

The only alternative that does not limit the compression ratio is the spark plug reach.

8036. Full-floating piston pins are those which allow motion between the pin and

A—the piston.
B—both the piston and the large end of the connecting rod.
C—both the piston and the small end of the connecting rod.

A full-floating piston pin is free to rotate in both the piston and the small end of the connecting rod.

Full-floating piston pins are usually a push fit in the piston. They are kept from damaging the cylinder walls as they move up and down by soft aluminum or brass plugs in the ends of the pin.

8037. The primary purpose in setting proper valve timing and overlap is to

A—permit the best possible charge of fuel/air mixture into the cylinders.
B—gain more thorough exhaust gas scavenging.
C—obtain the best volumetric efficiency and lower cylinder operating temperatures.

Valve overlap is the angular travel of the crankshaft during the time both the intake and exhaust valves are off their seats, and is used to increase the volumetric efficiency of the engine.

The exhaust valve remains open until after the piston has started down on the intake stroke to allow the maximum amount of burned exhaust gases to leave the cylinder.

The intake valve opens shortly before the piston reaches the top of its travel on the exhaust stroke. The

Answers
8032 [A] (A02) AMT-P
8033 [B] (A02) AMT-P
8034 [C] (A02) AMT-P
8035 [C] (A02) AMT-P
8036 [C] (A02) AMT-P
8037 [C] (A02) AMT-P

inertia of the exhaust gases leaving the cylinder when the intake valve opens, helps start the fresh fuel-air charge flowing into the cylinder.

By timing the valves and ignition to occur at the proper time, the mixture will not be burning as the piston is moving downward, and the cylinder walls will not become overheated.

8038. If the hot clearance is used to set the valves when the engine is cold, what will occur during operation of the engine?

A—The valves will open early and close early.
B—The valves will open late and close early.
C—The valves will open early and close late.

The cylinder head of an air-cooled engine expands much more than the pushrod. Because of this, air-cooled engines equipped with solid valve lifters (this applies primarily to radial engines) have a much larger valve clearance when the engine is hot than when it is cold.

If the valves are adjusted to the hot (running) clearance when the cylinder is cold, the clearance in the valve train will be too great when the engine is at its normal operating temperature.

The valves will open late and close early. The cam will have to turn farther to open the valve and the valve will close before the cam has turned to the normal valve-closing position.

8039. The purpose of two or more valve springs in aircraft engines is to

A—equalize side pressure on the valve stems.
B—eliminate valve spring surge.
C—equalize valve face loading.

Every mechanical device has a resonant frequency. If the valve is operating at the resonant frequency of the valve spring, the spring will lose its effectiveness and will surge, allowing the valve to float.

By using two or more valve springs wound with a different pitch and a different size wire, the resonant frequency of the springs will be different and there will be no engine RPM at which the valves will float.

8040. During overhaul, the disassembled parts of an engine are usually degreased with some form of mineral spirits solvent rather than water-mixed degreasers primarily because

A—solvent degreasers are much more effective.
B—water-mixed degreaser residues may cause engine oil contamination in the overhauled engine.
C—water-mixed degreasers cause corrosion.

Extreme care must be used if any water-mixed degreasing solutions containing caustic compounds of soap are used for cleaning engine parts. Such compounds, in addition to being potentially corrosive to aluminum and magnesium, may become impregnated in the pores of the metal and cause oil foaming when the engine is returned to service.

8041. Why does the smoothness of operation of an engine increase with a greater number of cylinders?

A—The power impulses are spaced closer together.
B—The power impulses are spaced farther apart.
C—The engine has larger counterbalance weights.

One of the main factors that affect the smoothness of operation of a reciprocating engine is the closeness with which the power impulses are spaced.

The greater the number of cylinders, the closer the power impulses are together and the smoother the engine will operate.

8042. Compression ratio is the ratio between the

A—piston travel on the compression stroke and on the intake stroke.
B—combustion chamber pressure on the combustion stroke and on the exhaust stroke.
C—cylinder volume with piston at bottom dead center and at top dead center.

The compression ratio of a reciprocating engine is the ratio of the volume of the cylinder with the piston at the bottom of its stroke to the volume of the cylinder with the piston at the top of its stroke.

8043. If the crankshaft runout readings on the dial indicator are plus .002 inch and minus .003 inch, the runout is

A—.005 inch.
B—plus .001 inch.
C—minus .001 inch.

Crankshaft run-out is measured by clamping a dial indicator to a solid part of the engine and placing the arm of the indicator against the part of the crankshaft where the run-out reading is to be measured.

Place the indicator at Zero with the arm against the crankshaft. Rotate the crankshaft for a complete revolution. The total run-out is the difference between the negative and the positive readings.

If the positive reading is +0.002 and the negative reading is -0.003, the total run-out is five thousandths of an inch (0.005 inch).

Answers

8038 [B] (A02) AMT-P
8039 [B] (A02) AMT-P
8040 [B] (A02) AMT-P
8041 [A] (A02) AMT-P
8042 [C] (A02) AMT-P
8043 [A] (A02) AMT-P

8044. (1) Cast iron piston rings may be used in chrome-plated cylinders.

(2) Chrome-plated rings may be used in plain steel cylinders.

Regarding the above statements,

A—only No. 1 is true.
B—neither No. 1 nor No. 2 is true.
C—both No. 1 and No. 2 are true.

Statement (1) is true. Only cast iron piston rings can be used in nitrided or chrome-plated cylinders.

Statement (2) is also true. Chrome plated rings can be used in plain steel cylinders.

8045. How is proper end-gap clearance on new piston rings assured during the overhaul of an engine?

A—By accurately measuring and matching the outside diameter of the rings with the inside diameter of the cylinders.
B—By using rings specified by the engine manufacturer.
C—By placing the rings in the cylinder and measuring the end-gap with a feeler gauge.

The end gap in piston rings is measured by placing the piston ring inside the cylinder and pushing it up with the top of the piston so that it is square in the cylinder bore and in line with the cylinder flange.

With the ring in this position, measure the distance between the two ends of the ring with a feeler gauge.

8046. The volume of a cylinder equals 70 cubic inches when the piston is at bottom center. When the piston is at the top of the cylinder, the volume equals 10 cubic inches. What is the compression ratio?

A—1:7.
B—7:10.
C—7:1.

The compression ratio of a reciprocating engine is the ratio of the volume of a cylinder with the piston at the bottom of its stroke to the volume of the cylinder with the piston at the top of its stroke.

If the cylinder has a volume of 70 cubic inches with the piston at the bottom of its stroke and 10 cubic inches with the piston at the top of its stroke, the compression ratio is 7:1.

8047. When cleaning aluminum and magnesium engine parts, it is inadvisable to soak them in solutions containing soap because

A—some of the soap will become impregnated in the surface of the material and subsequently cause engine oil contamination and foaming.
B—the soap can chemically alter the metals causing them to become more susceptible to corrosion.
C—the parts can be destroyed by dissimilar metal electrolytic action if they are placed together in the solution for more than a few minutes.

When cleaning aluminum and magnesium parts during engine overhaul, solutions containing soap should not be used, as it is very difficult to remove all traces of the soap.

When the engine is assembled and operating, heat will bring out any soap remaining on the surface or in the pores of the metal. This soap will contaminate the engine oil and cause severe foaming.

8048. What is the purpose of a power check on a reciprocating engine?

A—To check magneto drop.
B—To determine satisfactory performance.
C—To determine if the fuel/air mixture is adequate.

A power check of a reciprocating engine is a check to determine that the engine is developing the correct static RPM and manifold pressure.

The purpose of this check is to determine that the engine is performing satisfactorily.

8049. What will be the likely result if the piston ring gaps happen to be aligned when performing a differential-pressure compression check on a cylinder?

A—Little or no effect.
B—The rings will not be seated.
C—A worn or defective ring(s) indication.

The joints of the piston rings must be staggered around the circumference of the piston in which they are installed to reduce blowby.

If the gaps are not staggered, a differential compression check will give the indication of worn or defective rings.

8050. Which of the following will be caused by excessive valve clearance of a cylinder on a reciprocating aircraft engine?

A—Reduced valve overlap period.
B—Intake and exhaust valves will open early and close late.
C—A power increase by shortening the exhaust event.

Answers

8044 [C] (A02) AMT-P	8045 [C] (A02) AMT-P	8046 [C] (A02) AMT-P	8047 [A] (A02) AMT-P
8048 [B] (A03) AMT-P	8049 [C] (A03) AMT-P	8050 [A] (A03) AMT-P	

If both the intake and exhaust valves in a cylinder have excessive clearance, the valve overlap period will be reduced.

Valve overlap is the time between the end of the exhaust stroke and the beginning of the intake stroke when both valves are off of their seats.

If the intake valve clearance is too great, the intake valve will open late.

If the exhaust valve clearance is too great, the exhaust valve will close early.

Late opening of the intake valve and early closing of the exhaust valve, shorten the period of valve overlap.

8051. The floating control thermostat, used on some reciprocating engine installations, helps regulate oil temperature by

A—controlling oil flow through the oil cooler.
B—recirculating hot oil back through the sump.
C—controlling air flow through the oil cooler.

The floating-control thermostat controls the oil cooler air-exit door. It maintains the oil temperature within the desired limits by controlling the air flow through the oil cooler.

8052. Which of the following would indicate a general weak-engine condition when operated with a fixed-pitch propeller or test club?

A—Lower than normal static RPM, full throttle operation.
B—Manifold pressure lower at idle RPM than at static RPM.
C—Lower than normal manifold pressure for any given RPM.

The condition of an engine is shown by a full power check made with a fixed-pitch propeller or test club which furnishes a constant and a known load on the engine.

If the engine is not producing its full power, it will not produce the correct static RPM at full throttle. The static RPM will be too low, and low static RPM with a fixed propeller load indicates a "weak" engine.

8053. What is required by 14 CFR Part 43 Appendix D when performing an annual/100-hour inspection on a reciprocating engine aircraft?

A—Magneto timing check.
B—Cylinder compression check.
C—Valve clearance check.

The only alternative listed here that must be included in a 100-hour inspection as specified in 14 CFR Part 43, Appendix D, is the cylinder compression check.

A compression check, and specifically a differential compression check, tells much about the internal condition of the cylinders. It gives an indication of the seal provided by the valves and the condition of the piston rings.

8054. After spark plugs from an opposed engine have been serviced, in what position should they be reinstalled?

A—Next in firing order to the one from which they were removed.
B—Swapped bottom to top.
C—Next in firing order to the one from which they were removed and swapped bottom to top.

When spark plugs have been cleaned, gapped and tested, they should be installed in the cylinder next in firing order to the one from which they were removed, and they should be swapped from bottom to top.

8055. As the pressure is applied during a reciprocating engine compression check using a differential pressure tester, what would a movement of the propeller in the direction of engine rotation indicate?

A—The piston was on compression stroke.
B—The piston was on exhaust stroke.
C—The piston was positioned past top dead center.

When performing a differential compression check on a reciprocating engine, the piston of the cylinder being tested is placed on top center of the compression stroke and air is put into the cylinder.

If the air causes the propeller to turn in the direction of normal rotation, the piston is not on top dead center, but is slightly past top center.

8056. Excessive valve clearance results in the valves opening

A—late and closing early.
B—early and closing late.
C—late and closing late.

Excessive clearance in the valve train will cause the valves to open late (the cam will have to turn farther before the valve is opened) and close early (the valve will close before the cam rotates to the normal closing position).

Answers

8051 [C] (A03) AMT-P	8052 [A] (A03) AMT-P	8053 [B] (A03) 14 CFR 43	8054 [C] (A03) AMT-P
8055 [C] (A03) AMT-P	8056 [A] (A03) AMT-P		

8057. During routine inspection of a reciprocating engine, a deposit of small, bright, metallic particles which do not cling to the magnetic drain plug is discovered in the oil sump and on the surface of the oil filter. This condition

A—may be a result of abnormal plain type bearing wear and is cause for further investigation.
B—is probably a result of ring and cylinder wall wear and is cause for engine removal and/or overhaul.
C—is normal in engines utilizing plain type bearings and aluminum pistons and is not cause for alarm.

When you find any metal deposits in the lubricating-oil filters of an aircraft engine, you should investigate to find the source of the metal.

If the metal particles are not attracted by the magnetic drain plug, they are from either the plain bearings or the pistons.

8058. A characteristic of dyna-focal engine mounts as applied to aircraft reciprocating engines is that the

A—shock mounts eliminate the torsional flexing of the powerplant.
B—engine attaches to the shock mounts at the engine's center of gravity.
C—shock mounts point toward the engine's center of gravity.

Aircraft reciprocating engines are often mounted in a type of suspension called dynamic suspension, or dyna-focal engine mounts.

Dyna-focal mounts absorb the vibrations of the engine about the center of gravity of the engine-propeller combination and isolates these vibrations from the aircraft structure. The shock mounts all point toward the engine-propeller center of gravity.

8059. If metallic particles are found in the oil filter during an inspection,

A—it is an indication of normal engine wear unless the particles are nonferrous.
B—the cause should be identified and corrected before the aircraft is released for flight.
C—it is an indication of normal engine wear unless the deposit exceeds a specified amount.

Anytime metallic particles are found on the oil screen of an aircraft engine, their source and the cause for their being in the oil system must be determined and corrected before the aircraft is released for flight.

8060. If the oil pressure gauge fluctuates over a wide range from zero to normal operating pressure, the most likely cause is

A—low oil supply.
B—broken or weak pressure relief valve spring.
C—air lock in the scavenge pump intake.

Oil pressure fluctuation ranging from zero to the normal operating pressure is most likely caused by a low oil supply.

When the pump picks up oil, the pressure is normal, but when it draws air, the pressure drops to zero.

8061. What special procedure must be followed when adjusting the valves of an engine equipped with a floating cam ring?

A—Adjust valves when the engine is hot.
B—Adjust all exhaust valves before intake valves.
C—Eliminate cam bearing clearance when making valve adjustment.

Some large radial engines have floating cam rings.

A floating cam ring is held centered over its bearing by the forces exerted by the valve springs.

When checking the valve clearance on an engine equipped with a floating cam, the bearing clearance must be eliminated by depressing two valves on the opposite side of the engine from the valves being checked.

Depressing the valves removes the pressure of their valve spring from the cam allowing the cam ring to move tight against its bearing on the side where the valves are being checked.

8062. Which of the following is most likely to occur if an overhead valve engine is operated with inadequate valve clearances?

A—The valves will not seat positively during start and engine warmup.
B—The further decrease in valve clearance that occurs as engine temperatures increase will cause damage to the valve-operating mechanism.
C—The valves will remain closed for longer periods than specified by the engine manufacturer.

Overhead valves in an air-cooled engine have their smallest clearance when the engine is cold. This clearance opens up to several times the cold clearance when the engine is at its operating temperature.

If the valve clearance is too small, the valves will likely not seat positively when the engine is cold during start and engine warm-up.

Answers

8057 [A] (A03) AMT-P	8058 [C] (A03) AMT-P	8059 [B] (A03) AMT-P	8060 [A] (A03) AMT-P
8061 [C] (A03) AMT-P	8062 [A] (A03) AMT-P		

8063. Excessive valve clearances will cause the duration of valve opening to

A—increase for both intake and exhaust valves.
B—decrease for both intake and exhaust valves.
C—decrease for intake valves and increase for exhaust valves.

Excessive valve clearance will cause the valves to remain open for a shorter period of time than they would have with a normal clearance.

The cam must turn farther to open the valve and the valve will close before the cam has turned to the correct valve-closing position.

8064. What does valve overlap promote?

A—Lower intake manifold pressure and temperatures.
B—A backflow of gases across the cylinder.
C—Better scavenging and cooling characteristics.

Valve overlap is the portion of crankshaft rotation during which both the intake and the exhaust valves are off of their seats at the same time.

Adequate valve overlap increases the volumetric efficiency of the engine. It aids in the scavenging of the burned exhaust gases, and it gives the engine better cooling characteristics. This is done by ensuring that the fuel-air charge in the cylinder is rich enough for proper operation and not diluted with exhaust gases.

8065. At what speed must a crankshaft turn if each cylinder of a four-stroke cycle engine is to be fired 200 times a minute?

A—800 RPM.
B—1,600 RPM.
C—400 RPM.

Each cylinder in a four-stroke-cycle engine fires every other revolution of the crankshaft.

If a cylinder is to fire 200 times in one minute, the engine will have to be turning at 400 RPM.

8066. Engine crankshaft runout is usually checked

1. during engine overhaul.
2. during annual inspection.
3. after a "prop strike" or sudden engine stoppage.
4. during 100-hour inspection.

A—1, 3, and 4.
B—1 and 3.
C—1, 2 and 3.

Crankshaft run-out is checked to determine whether the crankshaft of a reciprocating engine is bent.

Crankshaft run-out is checked during each engine overhaul and after each sudden stoppage of the engine.

8067. Before attempting to start a radial engine that has been shut down for more than 30 minutes,

A—turn the propeller by hand three or four revolutions in the opposite direction of normal rotation to check for liquid lock.
B—turn the ignition switch on before energizing the starter.
C—turn the propeller by hand three to four revolutions in the normal direction of rotation to check for liquid lock.

There are some cylinders below the center line of a radial engine, and it is possible for oil to drain down, past the piston rings, into these lower cylinders while the engine is not operating.

When a radial engine has been shut down for a half hour or so, it should be checked for a liquid lock (oil in the lower cylinders) by pulling the propeller through in the direction of normal rotation by hand for at least two complete revolutions of the crankshaft.

If oil has collected in any of the lower cylinders, the spark plugs must be removed from these cylinders and all of the oil drained out.

8068. An engine misses in both the right and left positions of the magneto switch. The quickest method for locating the trouble is to

A—check for one or more cold cylinders.
B—perform a compression check.
C—check each spark plug.

If an engine misses on both magnetos, the quickest way to find the cylinder that is not firing is by running the engine at the RPM at which it misses the most consistently and by feeling the exhaust stack at the cylinder head.

The exhaust stack of the cylinder that is not firing will be much cooler than those of the cylinders that are firing normally.

8069. A hissing sound from the exhaust stacks when the propeller is being pulled through manually indicates

A—a cracked exhaust stack.
B—exhaust valve blow-by.
C—worn piston rings.

A hissing sound heard at the exhaust stacks when an aircraft engine is pulled through by hand is an indication that an exhaust valve is leaking. There is exhaust valve blow-by.

Answers

8063 [B] (A03) AMT-P	8064 [C] (A03) AMT-P	8065 [C] (A03) AMT-P	8066 [B] (A03) AMT-P
8067 [C] (A03) AMT-G	8068 [A] (A03) AMT-P	8069 [B] (A03) AMT-P	

8070. If the oil pressure of a cold engine is higher than at normal operating temperatures, the

A—oil system relief valve should be readjusted.
B—engine's lubrication system is probably operating normally.
C—oil dilution system should be turned on immediately.

Many large aircraft reciprocating engines have a compensating oil pressure relief valve that allows the oil pressure for cold oil to be considerably higher than it allows for warm oil. This higher pressure allows the thicker, higher viscosity oil to be forced through the engine bearings.

The plunger of the oil-pressure relief valve is held down by two springs when the oil is cold. However, when the oil warms up, a thermostatic valve opens and allows oil pressure to remove the force of one of the springs. For normal operation, only one spring holds the pressure relief valve on its seat.

8071. If an engine operates with a low oil pressure and a high oil temperature, the problem may be caused by a

A—leaking oil dilution valve.
B—sheared oil pump shaft.
C—clogged oil cooler annular jacket.

Some aircraft engines are equipped with an oil dilution system in which gasoline is put into the lubricating oil before the engine is shut down. Oil dilution is used in cold weather to make a cold engine easier to start.

If an oil dilution valve should leak and allow gasoline to flow into the oil supply during normal operation, it will cause the oil to be too thin (to have too low a viscosity) for normal operation. The oil pressure will drop and the oil temperature will go up.

8072. Which fuel/air mixture will result in the highest engine temperature (all other factors remaining constant)?

A—A mixture leaner than a rich best-power mixture of .085.
B—A mixture richer than a full-rich mixture of .087.
C—A mixture leaner than a manual lean mixture of .060.

Lean mixtures burn more slowly than rich mixtures and a mixture leaner than the manual-lean mixture of 0.060, or approximately 17:1, will possibly be burning as the gases are forced out past the exhaust valve. This will cause serious overheating of the exhaust valve.

8073. If an engine cylinder is to be removed, at what position in the cylinder should the piston be?

A—Bottom dead center.
B—Top dead center.
C—Halfway between top and bottom dead center.

When removing a cylinder from an aircraft engine, rotate the crankshaft until the piston is at top center on the compression stroke.

In this position, the pushrods can be most easily removed. The piston will be all the way out of the crankcase so the wrist pin can be slipped out to remove the piston with the cylinder.

8074. The horsepower developed in the cylinders of a reciprocating engine is known as the

A—shaft horsepower.
B—indicated horsepower.
C—brake horsepower.

Indicated horsepower (IHP) is the horsepower developed in the cylinders of a reciprocating engine without reference to friction losses. Shaft horsepower and brake horsepower are the actual usable powers that do include friction losses.

8075. Engine operating flexibility is the ability of the engine to

A—deliver maximum horsepower at a specific altitude.
B—meet exacting requirements of efficiency and low weight per horsepower ratio.
C—run smoothly and give the desired performance at all speeds.

One of the required characteristics of an aircraft engine is operating flexibility.

Operating flexibility is defined by the FAA as the ability of an engine to run smoothly and to give the desired performance at all speeds.

8076. Standard aircraft cylinder oversizes usually range from 0.010 inch to 0.030 inch. Oversize on automobile engine cylinders may range up to 0.100 inch. This is because aircraft engine cylinders

A—have more limited cooling capacity.
B—have relatively thin walls and may be nitrided.
C—operate at high temperatures.

Thin-walled aircraft-engine cylinders may be ground oversize, but the amount they may be ground is much less than is allowed for the much thicker-walled cylinders normally used in automobile engines.

Some aircraft-engine cylinders may not be re-bored at all. The engine manufacturer's recommendations must be followed in detail regarding any re-boring operation.

Answers

8070 [B] (A03) AMT-P	8071 [A] (A03) AMT-P	8072 [C] (A03) AMT-P	8073 [B] (A03) AMT-P
8074 [B] (A03) AMT-P	8075 [C] (A03) AMT-P	8076 [B] (A03) AMT-P	

8077. If the ignition switch is moved from BOTH to either LEFT or RIGHT during an engine ground check, normal operation is usually indicated by a

A—large drop in RPM.
B—momentary interruption of both ignition systems.
C—slight drop in RPM.

All certificated aircraft reciprocating engines have dual ignition. During normal operation, both ignition systems are operating.

During a magneto check, the engine is operated at the speed specified by the engine manufacturer and the ignition system is switched from BOTH magnetos to each magneto separately.

When the engine operates on a single magneto, the fuel-air mixture in the cylinder is ignited at one point only causing a slight drop in engine power. The RPM will drop slightly.

8078. During ground check an engine is found to be rough-running, the magneto drop is normal, and the manifold pressure is higher than normal for any given RPM. The trouble may be caused by

A—several spark plugs fouled on different cylinders.
B—a leak in the intake manifold.
C—a dead cylinder.

A dead cylinder will cause an engine to run rough. Because the throttle will have to be opened farther to get the same RPM, the manifold pressure will be higher than it would on an engine with all the cylinders firing.

A dead cylinder will not show up on a magneto check.

8079. What is the best indication of worn valve guides?

A—High oil consumption.
B—Low compression.
C—Low oil pressure.

High oil consumption is the only alternative that would indicate worn valve guides.

When the valve guide wears, oil from the rocker box flows down the valve stem and is burned.

8080. By use of a differential pressure compression tester, it is determined that the No. 3 cylinder of a nine-cylinder radial engine will not hold pressure after the crankshaft has been rotated 260° from top dead center compression stroke No. 1 cylinder. How can this indication usually be interpreted?

A—A normal indication.
B—Exhaust valve blow-by.
C—A damaged exhaust valve or insufficient exhaust valve clearance.

In a nine-cylinder radial engine, each cylinder fires 80° of crankshaft rotation after the cylinder before it, in firing order.

When the crankshaft is rotated 260° after the piston in cylinder number one is at top dead center on its compression stroke, the piston in cylinders 7 and 8 are near the top of their strokes. The piston in cylinder 3 is near the bottom of its power stroke and its exhaust valve is open.

It is normal for a cylinder not to hold air pressure when its piston is near the bottom of its power stroke and its exhaust valve is open.

8081. When does valve overlap occur in the operation of an aircraft reciprocating engine?

A—At the end of the exhaust stroke and the beginning of the intake stroke.
B—At the end of the power stroke and the beginning of the exhaust stroke.
C—At the end of the compression stroke and the beginning of the power stroke.

Both the intake and exhaust valve are open at the same time, only during the period of valve overlap.

Valve overlap occurs at the end of the exhaust stroke and the beginning of the intake stroke.

The intake valve opens a few degrees of crankshaft rotation before the piston reaches the top of the exhaust stroke. The exhaust valve remains open until the piston has moved down a few degrees of crankshaft rotation on the intake stroke.

8082. What is an advantage of using metallic-sodium filled exhaust valves in aircraft reciprocating engines?

A—Increased strength and resistance to cracking.
B—Reduced valve operating temperatures.
C—Greater resistance to deterioration at high valve temperatures.

Some aircraft engine exhaust valves are hollow and are partially filled with metallic sodium. When the engine is operating, the sodium melts and as the valve opens and closes, the molten sodium sloshes back and forth in the valve. When it is in the head, it absorbs heat. When it is in the stem, it transfers this heat to the valve guides.

Sodium-filled valves reduce the valve operating temperature.

Answers
8077 [C] (A03) AMT-P
8078 [C] (A03) AMT-P
8079 [A] (A03) AMT-P
8080 [A] (A03) AMT-P
8081 [A] (A03) AMT-P
8082 [B] (A03) AMT-P

8083. Valve clearance changes on opposed-type engines using hydraulic lifters are accomplished by

A—rocker arm adjustment.
B—rocker arm replacement.
C—push rod replacement.

When assembling an opposed engine equipped with hydraulic valve lifters, if the valve clearance is not within the allowable limits, install a pushrod of a slightly different length.

8084. What is likely to occur if a reciprocating engine is operated at high power settings before it is properly warmed up?

A—Oil starvation of bearings and other parts.
B—Excessive thinning of the engine oil.
C—Accelerated oil breakdown and oxidation.

No aircraft engine should be operated at high power settings before it is properly warmed up and the oil is warm enough to flow freely through all the passages. High power operation with cold oil can cause oil starvation to the bearings.

8085. An increase in manifold pressure with a constant RPM will cause the bearing load in an engine to

A—decrease.
B—remain relatively constant.
C—increase.

The cylinder pressure applied to the crankshaft through the connecting rod bearings is determined by the compression ratio of the engine and the manifold pressure.

If the manifold pressure for a given RPM is increased, the bearing load imposed on the crankshaft will increase.

8086. Direct mechanical push-pull carburetor heat control linkages should normally be adjusted so that the stop located on the diverter valve will be contacted

A—before the stop at the control lever is reached in both HOT and COLD positions.
B—before the stop at the control lever is reached in the HOT position and after the stop at the control lever is reached in the COLD position.
C—after the stop at the control lever is reached in both HOT and COLD positions.

When rigging any engine control in an aircraft, the stop on the component being actuated must be contacted before the stop in the cockpit.

The control linkage has enough spring-back in both directions that after the stop on the diverter valve is contacted, the control can be moved in both directions until it contacts the stop in the cockpit.

When the control is released, it will spring back a few degrees.

8087. Reduced air density at high altitude has a decided effect on carburetion, resulting in a reduction of engine power by

A—excessively enriching the fuel/air mixture.
B—excessively leaning the fuel/air mixture.
C—reducing fuel vaporization.

An aircraft engine produces power by converting the chemical energy in the fuel into heat energy as the fuel-air mixture is burned inside the engine cylinders.

The efficiency of this energy interchange is determined by the ratio between weight of the air and the weight of the fuel in the mixture.

The air at high altitude is less dense (weighs less) than the air at sea level, and the fuel metered into the same volume of air will cause the fuel-air mixture at high altitude to become excessively rich. There will be too many pounds of fuel per pound of air for the most efficient production of power.

8088. Increased water vapor (higher relative humidity) in the incoming air to a reciprocating engine will normally result in which of the following?

A—Decreased engine power at a constant RPM and manifold pressure.
B—Increased power output due to increased volumetric efficiency.
C—A leaning effect on engines which use non-automatic carburetors.

The amount of energy released by a burning fuel-air mixture is determined by the weight of both the fuel and the air in the mixture.

Water vapor weighs only about 5/8 as much as dry air, and when an engine takes in air with a high relative humidity, it produces less power at the same RPM and manifold pressure than it would produce if it were taking in dry air.

8089. (1) Preignition is caused by improper ignition timing.

(2) Detonation occurs when an area of the combustion chamber becomes incandescent and ignites the fuel/air mixture in advance of normal timed ignition.

Regarding the above statements,

A—only No. 1 is true.
B—both No. 1 and No. 2 are true.
C—neither No. 1 nor No. 2 is true.

Answers

8083 [C] (A03) AMT-P	8084 [A] (A03) AMT-P	8085 [C] (A04) AMT-P	8086 [A] (A04) AMT-P
8087 [A] (A04) AMT-P	8088 [A] (A04) AMT-P	8089 [C] (A04) AMT-P	

Statement (1) is not true. Preignition is the ignition of the fuel-air mixture before normal ignition is timed to occur. It is caused by incandescent objects in the cylinder. Detonation produces enough heat in a cylinder that carbon particles can become incandescent and ignite the mixture early.

Statement (2) is not true. Detonation is the spontaneous combustion of the unburned charge ahead of the flame front after ignition has occurred.

Preignition can cause the fuel-air mixture to burn in the cylinder long enough to heat the unburned mixture to its critical temperature. At this point, it explodes rather than burns. This instantaneous release of energy can overheat the cylinder producing enough pressure to damage the piston and connecting rod.

8090. Which of the following engine servicing operations generally requires engine pre-oiling prior to starting the engine?

A—Engine oil and filter change.
B—Engine installation.
C—Replacement of oil lines.

When a new or freshly overhauled engine is installed in an aircraft, it must be pre-oiled.

The oil tank is filled, and oil is pumped through all the passages until pressure registers on the oil pressure gauge in the cockpit.

Pre-oiling ensures that all the bearings will be adequately lubricated before the oil pump begins to pump oil through the system normally.

8091. During the inspection of an engine control system in which push-pull control rods are used, the threaded rod ends should

A—not be adjusted in length for rigging purposes because the rod ends have been properly positioned and staked during manufacture.
B—be checked for thread engagement of at least two threads but not more than four threads.
C—be checked for the amount of thread engagement by means of the inspection holes.

When installing push-pull control rods in an aircraft, you can determine that the rod end is screwed onto the rod properly by trying to pass a piece of safety wire through the inspection hole in the rod. If the rod end is screwed into the rod far enough to cover the hole, there are enough threads engaged to give the connection the full strength required.

8092. Which of the following conditions would most likely lead to detonation?

A—Late ignition timing.
B—Use of fuel with too high an octane rating.
C—Use of fuel with too low an octane rating.

Detonation is an uncontrolled burning of the fuel inside the engine cylinders. The fuel-air mixture actually explodes, rather than burning evenly as it should.

The octane rating of a fuel is a measure of its detonation resistance. If a fuel with too low an octane rating is used, the fuel-air mixture is likely to detonate when the engine is developing full power.

8093. An unsupercharged aircraft reciprocating engine, operated at full throttle from sea level, to 10,000 feet, provided the RPM is unchanged, will

A—lose power due to the reduced volume of air drawn into the cylinders.
B—produce constant power due to the same volume of air drawn into the cylinders.
C—lose power due to the reduced density of the air drawn into the cylinders.

The power produced by an aircraft reciprocating engine is determined by the weight of the air that is mixed with the fuel and burned.

At altitude, the density of the air (its weight per unit volume) is less than it is at sea level.

Therefore, for the same RPM, a unsupercharged engine will take in less weight of air to combine with the fuel and the engine will lose power.

8094. Which of the following would most likely cause a reciprocating engine to backfire through the induction system at low RPM operation?

A—Idle mixture too rich.
B—Clogged derichment valve.
C—Lean mixture.

A lean fuel-air mixture burns slower than either a rich or a chemically-correct mixture. There is a possibility that a lean mixture will still be burning as it is pushed out through the exhaust valve.

During the time of valve overlap, when both the intake and the exhaust valves are open, the burning exhaust gases can ignite the fresh fuel-air charge being taken into the cylinder through the intake valve. This can cause a backfire through the induction system.

Answers

8090 [B] (A04) AMT-P	8091 [C] (A04) AMT-P	8092 [C] (A04) AMT-P	8093 [C] (A04) AMT-P
8094 [C] (A04) AMT-P			

8095. How may it be determined that a reciprocating engine with a dry sump is pre-oiled sufficiently?

A—The engine oil pressure gauge will indicate normal oil pressure.
B—Oil will flow from the engine return line or indicator port.
C—When the quantity of oil specified by the manufacturer has been pumped into the engine.

When pre-oiling a dry-sump reciprocating engine, you know there is oil in all the passages when oil flows from the engine return line or from the port to which the oil pressure gauge is connected.

8096. What is the basic operational sequence for reducing the power output of an engine equipped with a constant-speed propeller?

A—Reduce the RPM, then the manifold pressure.
B—Reduce the manifold pressure, then retard the throttle to obtain the correct RPM.
C—Reduce the manifold pressure, then the RPM.

When reducing the power of an engine equipped with a constant-speed propeller, it is important that the manifold pressure be reduced by retarding the throttle before the RPM is reduced with the propeller pitch control.

If the wrong sequence is used, the high manifold pressure and the low RPM can produce cylinder pressures high enough to seriously damage the engine.

8097. Which statement pertaining to fuel/air ratios is true?

A—The mixture ratio which gives the best power is richer than the mixture ratio which gives maximum economy.
B—A rich mixture is faster burning than a normal mixture.
C—The mixture ratio which gives maximum economy may also be designated as best power mixture.

The fuel-air mixture used for the engine to produce its best power is richer (there is more fuel for the air) than a mixture that gives the maximum economy.

The best power mixture is about a 12:1 mixture (12 parts of air to one part of fuel), and the maximum economy mixture is about 16:1.

8098. Backfiring through the carburetor generally results from the use of

A—an excessively lean mixture.
B—excessively atomized fuel.
C—an excessively rich mixture.

Backfiring through the carburetor is often caused by the use of an extremely lean mixture.

A lean mixture burns slowly. If it is still burning when the intake valve opens, the burning mixture will ignite the fresh fuel-air charge and cause a backfire in the induction system.

8099. Which of these conditions will cause an engine to have an increased tendency to detonate?

1. High manifold pressure.
2. High intake air temperature.
3. Engine overheated.
4. Late ignition timing.

A—1, 4.
B—1, 2, 3.
C—1, 2, 3, 4.

Detonation occurs when the fuel-air mixture burning in a cylinder reaches its critical pressure and temperature.

Detonation may be caused by the high pressure and temperature resulting from high manifold pressure, high intake air temperature, or an overheated engine.

Late ignition timing reduces engine power but it does not cause detonation.

8100. When will small induction system air leaks have the most noticeable effect on engine operation?

A—At high RPM.
B—At maximum continuous and takeoff power settings.
C—At low RPM.

A small induction-system air leak will have the most noticeable effect on engine operation when the engine is operating at low RPM.

At the low engine speed, the volume of air entering the cylinders is small. Because of this, the additional air coming in through the leak makes an appreciable change in the fuel-air mixture ratio.

At higher RPMs, so much air is being taken into the cylinders that the amount that leaks into the system does not change the ratio enough to make a big difference.

8101. To reduce the power output of an engine equipped with a constant-speed propeller and operating near maximum BMEP, the

A—manifold pressure is reduced with the throttle control before the RPM is reduced with the propeller control.
B—manifold pressure is reduced with the propeller control before the RPM is reduced with the throttle control.
C—RPM is reduced with the propeller control before the manifold pressure is reduced with the throttle control.

Answers

8095 [B] (A04) AMT-P	8096 [C] (A04) AMT-P	8097 [A] (A04) AMT-P	8098 [A] (A04) AMT-P
8099 [B] (A04) AMT-P	8100 [C] (A04) AMT-P	8101 [A] (A04) AMT-P	

When changing the power setting of an engine equipped with a constant-speed propeller, it is important that the manifold pressure be reduced by retarding the throttle before the RPM is reduced with the propeller pitch control.

If the wrong sequence is used, the high manifold pressure and the low RPM can produce cylinder pressures high enough to seriously damage the engine.

8102. One of the best indicators of reciprocating engine combustion chamber problems is

A—excessive engine vibration.
B—starting difficulties.
C—spark plug condition.

The condition of the spark plugs taken from the cylinders of a reciprocating engine is a good indicator of the condition of the combustion chamber of the engine.

Spark plugs can show when detonation has been occurring, and they can show up an excessively worn valve guide and induction system filter leaks.

8103. What could cause excessive pressure buildup in the crankcase of a reciprocating engine?

A—Plugged crankcase breather.
B—Improper warmup operation.
C—An excessive quantity of oil.

The crankcase of a reciprocating engine is vented to the outside air through a breather pipe.

If the breather should ever become plugged, the crankcase pressure can build up to a point at which the oil consumption increases drastically. It is possible that the oil can then be forced into engine accessories damaging them.

8104. Excessive valve clearance in a piston engine

A—increases valve overlap.
B—increases valve opening time.
C—decreases valve overlap.

Excessive valve clearance in a reciprocating engine will decrease the valve overlap.

The intake valve will open late and the exhaust valve will close early.

8105. To what altitude will a turbo charged engine maintain sea level pressure?

A—Critical altitude.
B—Service ceiling.
C—Pressure altitude.

The critical altitude of a turbocharged aircraft engine is the altitude above which the turbocharger can no longer produce sea level manifold pressure and the engine cannot maintain its rated horsepower.

8106. If air is heard coming from the crankcase breather or oil filler during a differential compression check, what is this an indication of?

A—Exhaust valve leakage.
B—Intake valve leakage.
C—Piston ring leakage.

A hissing sound heard at the crankcase breather during a differential compression check is caused by air leaking past the piston rings.

Exhaust valve leakage is heard at the exhaust stack or muffler and intake valve leakage is heard at the carburetor inlet.

8107. One cause of afterfiring in an aircraft engine is

A—sticking intake valves.
B—an excessively lean mixture.
C—an excessively rich mixture.

After-firing, or torching, is the burning of the fuel-air mixture in the exhaust manifold after the mixture has passed through the exhaust valve.

After-firing is usually caused by operation with an excessively rich mixture, such as would be caused by overpriming, improper use of the mixture control when starting, or by poor ignition.

8108. At what point in an axial-flow turbojet engine will the highest gas pressures occur?

A—At the turbine entrance.
B—Within the burner section.
C—At the compressor outlet.

The highest gas pressure in an axial-flow turbojet engine occurs at the outlet of the compressor.

The compressor outlet is the same as the burner inlet.

Answers

8102 [C] (A04) AMT-P	8103 [A] (A04) AMT-P	8104 [C] (A04) AMT-P	8105 [A] (A04) AMT-P
8106 [C] (A04) AMT-P	8107 [C] (A04) AMT-P	8108 [C] (B01) AMT-P	

8109. One function of the nozzle diaphragm in a turbine engine is to?

A—Decrease the velocity of exhaust gases.
B—Center the fuel spray in the combustion chamber.
C—Direct the flow of gases to strike the turbine blades at the desired angle.

One of the functions of the nozzle diaphragm in a turbine engine is to deflect the gases to a specific angle in the direction of the turbine wheel rotation.

8110. What is the profile of a turbine engine compressor blade?

A—The leading edge of the blade.
B—A cutout that reduces blade tip thickness.
C—The curvature of the blade root.

"Profile" in a rotor blade of a turbine-engine compressor is a reduction in the thickness of the blade tip.

Profiles prevent serious damage to the blade or the housing if the blade should contact the compressor housing.

8111. The fan rotational speed of a dual axial compressor forward fan engine is the same as the

A—low-pressure compressor.
B—forward turbine wheel.
C—high-pressure compressor.

The fan is a portion of the low-pressure compressor of a dual axial-flow compressor. The rotor blades in the fan section are long enough that the air they move passes around the outside of the gas generator portion of the engine.

The fan rotational speed is the same as that of the low-pressure compressor.

8112. The abbreviation "P" with subscript t7 used in turbine engine terminology means

A—the total inlet pressure.
B—pressure and temperature at station No. 7.
C—the total pressure at station No. 7.

Pressures in a turbine engine are identified according to their type and to the location at which they are measured.

P_{t7} is the total pressure (the pressure a body of moving fluid has when its movement is stopped) measured at station 7, the turbine discharge.

8113. The blending of blades and vanes in a turbine engine

A—is usually accomplished only at engine overhaul.
B—should be performed parallel to the length of the blade using smooth contours to minimize stress points.
C—may sometimes be accomplished with the engine installed, ordinarily using power tools.

Blending is a method of hand recontouring damaged compressor blades and vanes using small files, emery cloth and honing stones. Blending is done parallel to the length of the blade to minimize stress concentrations and to restore a smooth a shape to the surface.

8114. What turbine engine section provides for proper mixing of the fuel and air?

A—Combustion section.
B—Compressor section.
C—Diffuser section.

In a turbojet engine, compressed air from the compressor is directed into the combustion section.

Fuel is sprayed from nozzles in the combustion section where it mixes with the air and burns.

8115. In a gas turbine engine, combustion occurs at a constant

A—volume.
B—pressure.
C—density.

The energy exchange cycle used in a turbojet engine is the Brayton cycle which is a constant-pressure cycle. The pressure of the air remains relatively constant as the energy from the burning fuel is added to it.

Since the pressure remains constant as the volume of the gas increases, its velocity increases.

8116. Which statement is true regarding jet engines?

A—At the lower engine speeds, thrust increases rapidly with small increases in RPM.
B—At the higher engine speeds, thrust increases rapidly with small increases in RPM.
C—The thrust delivered per pound of air consumed is less at high altitude than at low altitude.

The relationship between thrust and RPM is such that the amount of thrust increases rapidly as the engine speed increases.

A small change in speed at low RPM will not produce nearly as much increase in thrust as the same amount of change at high RPM.

Answers

8109 [C] (B01) AMT-P	8110 [B] (B01) AMT-P	8111 [A] (B01) AMT-P	8112 [C] (B01) AMT-P
8113 [B] (B01) AMT-P	8114 [A] (B01) AMT-P	8115 [B] (B01) AMT-P	8116 [B] (B01) AMT-P

8117. Some high-volume turboprop and turbojet engines are equipped with two-spool or split compressors. When these engines are operated at high altitudes, the

A—low-pressure rotor will increase in speed as the compressor load decreases in the lower density air.
B—throttle must be retarded to prevent overspeeding of the high-pressure rotor due to the lower density air.
C—low-pressure rotor will decrease in speed as the compressor load decreases in the lower density air.

The high-pressure rotor of a two-spool (split-spool) compressor is governed for speed, but the low-pressure rotor is free to operate at its own best speed.

As the air density decreases at altitude, the compressor load decreases and the low-pressure rotor increases its speed.

8118. Turbine nozzle diaphragms located on the upstream side of each turbine wheel, are used in the gas turbine engine to

A—decrease the velocity of the heated gases flowing past this point.
B—direct the flow of gases parallel to the vertical line of the turbine blades.
C—increase the velocity of the heated gases flowing past this point.

One of the functions of the nozzle diaphragm in a turbojet engine is to increase the velocity of the heated gases flowing through it.

In speeding up this gas, a portion of the heat and pressure energy is turned into velocity energy, which is converted to mechanical energy by the turbine rotor blades.

8119. Where is the highest gas pressure in a turbojet engine?

A—At the outlet of the tailpipe section.
B—At the entrance of the turbine section.
C—In the entrance of the burner section.

The highest gas pressure in a turbojet engine occurs at the outlet of the compressor.

The compressor outlet is the same as the entrance of the burner section.

8120. An exhaust cone placed aft of the turbine in a jet engine will cause the pressure in the first part of the exhaust duct to

A—increase and the velocity to decrease.
B—increase and the velocity to increase.
C—decrease and the velocity to increase.

The exhaust cone on a turbojet engine forms a divergent duct, which increases the pressure of the exiting gases and decreases their velocity.

8121. What is the function of the stator vane assembly at the discharge end of a typical axial-flow compressor?

A—To straighten airflow to eliminate turbulence.
B—To direct the flow of gases into the combustion chambers.
C—To increase air swirling motion into the combustion chambers.

The stator vanes at the discharge of an axial-flow compressor are called the straightening vanes. They are used to straighten the airflow to eliminate turbulence.

8122. The turbine section of a jet engine

A—increases air velocity to generate thrust forces.
B—utilizes heat energy to expand and accelerate the incoming gas flow.
C—drives the compressor section.

One of the major functions of the turbine section in a turbojet or turbofan engine is to drive the compressor.

8123. When starting a turbine engine,

A—a hot start is indicated if the exhaust gas temperature exceeds specified limits.
B—an excessively lean mixture is likely to cause a hot start.
C—release the starter switch as soon as indication of light-off occurs.

A hot start is indicated if the engine starts but the exhaust gas temperature exceeds specified limits.

8124. In the dual axial-flow or twin spool compressor system, the first stage turbine drives the

A—N_1 and N_2 compressors.
B—N_2 compressor.
C—N_1 compressor.

The first-stage turbine in a twin-spool turbojet engine drives the high-pressure compressor.

This is the second stage of compression and is called the N_2 compressor.

Answers

8117 [A] (B01) AMT-P	8118 [C] (B01) AMT-P	8119 [C] (B01) AMT-P	8120 [A] (B01) AMT-P
8121 [A] (B01) AMT-P	8122 [C] (B01) AMT-P	8123 [A] (B01) AMT-G	8124 [B] (B01) AMT-P

8125. During inspection, turbine engine components exposed to high temperatures may only be marked with such materials as allowed by the manufacturer. These materials generally include

1. layout dye.
2. commercial felt tip marker.
3. wax or grease pencil.
4. chalk.
5. graphite lead pencil.

A—1, 2, and 4.
B—1, 3, and 4.
C—2, 4, and 5.

Materials used to mark the components in the hot section of a turbine engine must be carefully chosen so that they do not contaminate the structure of the metal when it is exposed to high temperatures. Materials such as lead pencils and wax or grease pencils which contain carbon, will cause granular embrittlement and lead to cracking, must not be used.

Layout dye, commercial felt tip markers, and chalk are normally allowed.

8126. When starting a turbine engine, a hung start is indicated if the engine

A—exhaust gas temperature exceeds specified limits.
B—fails to reach idle RPM.
C—RPM exceeds specified operating speed.

A hung start in a turbine engine is a start in which the engine lights off, but is unable to accelerate to a speed high enough to keep running without help from the starter.

8127. What are the two basic elements of the turbine section in a turbine engine?

A—Impeller and diffuser.
B—Hot and cold.
C—Stator and rotor.

The turbine section of a gas turbine engine consists of two basic elements, the stator and the rotor.

8128. The function of the exhaust cone assembly of a turbine engine is to

A—collect the exhaust gases and act as a noise suppressor.
B—swirl and collect the exhaust gases into a single exhaust jet.
C—straighten and collect the exhaust gases into a solid exhaust jet.

The primary function of an exhaust cone assembly used on a turbine engine is to collect the exhaust gases after they pass through the turbine and convert them into a solid, high-velocity exhaust jet.

8129. What are the two functional elements in a centrifugal compressor?

A—Turbine and compressor.
B—Bucket and expander.
C—Impeller and diffuser.

A centrifugal compressor used in a gas turbine engine has two basic functional elements, the impeller and the diffuser.

The impeller adds energy to the air flowing through the engine and speeds it up. The diffuser slows the air down and increases its pressure.

8130. What must be done after the fuel control unit has been replaced on an aircraft gas turbine engine?

A—Perform a full power engine run to check fuel flow.
B—Recalibrate the fuel nozzles.
C—Retrim the engine.

After a fuel control has been replaced on a turbine engine, the engine must be re-trimmed.

Trimming a gas turbine engine consists of adjusting the fuel control to give the engine the correct idle and maximum RPM.

8131. If, during inspection at engine overhaul, ball or roller bearings are found to have magnetism but otherwise have no defects, they

A—cannot be used again.
B—are in an acceptable service condition.
C—must be degaussed before use.

Ball and roller bearings that are in good condition but are shown to have magnetism in them may have the magnetism removed with a suitable degausser. If the bearings are allowed to remain in their magnetized state they will be damaged by the foreign ferrous particles they attract.

8132. A turbine engine compressor which contains vanes on both sides of the impeller is a

A—double entry centrifugal compressor.
B—double entry axial-flow compressor.
C—single entry axial-flow compressor.

A centrifugal compressor with vanes on both sides of the impeller is called a double-entry centrifugal compressor.

Answers

8125 [A] (B01) AMT-P	8126 [B] (B01) AMT-P	8127 [C] (B01) AMT-P	8128 [C] (B01) AMT-P
8129 [C] (B01) AMT-P	8130 [C] (B01) AMT-P	8131 [C] (B01) AMT-P	8132 [A] (B01) AMT-P

8133. What is the first engine instrument indication of a successful start of a turbine engine?

A—A rise in the engine fuel flow.
B—A rise in oil pressure.
C—A rise in the exhaust gas temperature.

When starting a gas turbine engine, the first indication of a successful start is a sudden rise in the exhaust-gas temperature.

8134. Some engine manufacturers of twin spool gas turbine engines identify turbine discharge pressure in their maintenance manuals as

A—Pt7.
B—Pt2.
C—Tt7.

Turbine-discharge total pressure is identified in maintenance manuals by the abbreviation P_{t7}. P_{t2} is compressor-inlet total pressure, and T_{t7} is turbine-discharge total temperature.

8135. Who establishes the recommended operating time between overhauls (TBO) of a turbine engine used in general aviation?

A—The engine manufacturer.
B—The operator (utilizing manufacturer data and trend analysis) working in conjunction with the FAA.
C—The FAA.

The engine manufacturer establishes the recommended TBO of a turbine engine, and these times are approved by the FAA.

8136. The basic gas turbine engine is divided into two main sections: the cold section and the hot section.

(1) The cold section includes the engine inlet, compressor, and turbine sections.

(2) The hot section includes the combustor, diffuser, and exhaust sections.

Regarding the above statements,

A—only No. 1 is true.
B—only No. 2 is true.
C—neither No. 1 nor No. 2 is true.

Statement (1) is not true. The cold section of a turbine engine does not include the turbine sections.

Statement (2) is not true because the diffuser is part of the cold section of the engine.

8137. (1) Welding and straightening of turbine engine rotating airfoils does not require special equipment.

(2) Welding and straightening of turbine engine rotating airfoils is commonly recommended by the manufacturer.

Regarding the above statements,

A—only No. 1 is true.
B—only No. 2 is true.
C—neither No. 1 nor No. 2 is true.

Neither statement is true. Welding and straightening of rotating airfoils in a gas turbine engine require special equipment. Quite often, neither procedure is authorized by the engine manufacturer.

8138. Turbine engine components exposed to high temperatures generally may NOT be marked with

1. layout dye.
2. commercial felt tip marker.
3. wax or grease pencil.
4. chalk.
5. graphite lead pencil.

A—1, 2, and 3.
B—3 and 5.
C—4 and 5.

Materials used to mark the components in the hot section of a turbine engine must be carefully chosen so that they do not contaminate the structure of the metal when it is exposed to high temperatures. Materials such as lead pencils and wax or grease pencils which contain carbon will cause granular embrittlement and lead to cracking must not be used.

8139. Who establishes mandatory replacement times for critical components of turbine engines?

A—The FAA.
B—The operator working in conjunction with the FAA.
C—The engine manufacturer.

The Instructions for Continued Airworthiness, prepared by the engine manufacturer and approved by the FAA, contain the mandatory replacement times for critical components of the engines to which the instructions apply.

Answers

8133 [C] (B01) AMT-P	8134 [A] (B01) AMT-P	8135 [A] (B01) 14 CFR 33	8136 [C] (B01) AMT-P
8137 [C] (B01) AMT-P	8138 [B] (B01) AMT-P	8139 [C] (B01) 14 CFR 33.4	

8140. Main bearing oil seals used with turbine engines are usually what type(s)?

A—Labyrinth and/or carbon rubbing.
B—Teflon and synthetic rubber.
C—Labyrinth and/or silicone rubber.

The bearing housing of a turbine engine usually contains seals to prevent loss of oil into the gas path. Oil seal are usually of the labyrinth or carbon rubbing type.

8141. How does a dual axial-flow compressor improve the efficiency of a turbojet engine?

A—More turbine wheels can be used.
B—Higher compression ratios can be obtained.
C—The velocity of the air entering the combustion chamber is increased.

The use of two axial-flow compressors turning at different speeds allows higher compression ratios to be obtained without the danger of compressor stall.

8142. Three types of turbine blades are

A—reaction, converging, and diverging.
B—impulse, reaction, and impulse-reaction.
C—impulse, vector, and impulse-vector.

Three basic types of turbine blades used in gas turbine engines are impulse, reaction and impulse-reaction.

8143. Which statements are true regarding aircraft engine propulsion?

1. An engine driven propeller imparts a relatively small amount of acceleration to a large mass of air.
2. Turbojet and turbofan engines impart a relatively large amount of acceleration to a smaller mass of air.
3. In modern turboprop engines, nearly 50 percent of the exhaust gas energy is extracted by turbines to drive the propeller and compressor with the rest providing exhaust thrust.

A—1, 2, 3.
B—1, 2.
C—1, 3.

The basic difference in the thrust produced by a turbojet or turbofan engine and that produced by a turboprop engine is in the mass of air moved and amount of acceleration imparted to it.

An engine-driven propeller imparts a relatively small amount of acceleration to a large mass of air, and a turbojet or turbofan engine imparts a greater amount of acceleration to a smaller mass of air.

Almost all of the useful heat energy in a turboprop engine is used to drive the compressor and the propeller, and very little is used to provide exhaust thrust.

8144. An advantage of the axial-flow compressor is its

A—low starting power requirements.
B—low weight.
C—high peak efficiency.

An axial-flow compressor has an advantage over a centrifugal-flow compressor in that it has a higher peak efficiency.

8145. What is one purpose of the stator blades in the compressor section of a turbine engine?

A—Stabilize the pressure of the airflow.
B—Control the direction of the airflow.
C—Increase the velocity of the airflow.

The stator blades in an axial-flow compressor convert the high-velocity energy of the air into pressure energy.

They also direct the airflow from each of the rotor stages to obtain the maximum possible blade efficiency.

8146. What is the purpose of the diffuser section in a turbine engine?

A—To increase pressure and reduce velocity.
B—To convert pressure to velocity.
C—To reduce pressure and increase velocity.

The diffuser section in a centrifugal-flow turbojet engine reduces the velocity of the air as it leaves the compressor and increases its pressure.

8147. Where do stress rupture cracks usually appear on turbine blades?

A—Across the blade root, parallel to the fir tree.
B—Along the leading edge, parallel to the edge.
C—Across the leading or trailing edge at a right angle to the edge length.

Stress rupture cracks usually appear as tiny hairline cracks on or across the leading or trailing edge of the turbine blades at right angles to the edge length of the blade.

8148. In which type of turbine engine combustion chamber is the case and liner removed and installed as one unit during routine maintenance?

A—Can.
B—Can annular.
C—Annular.

Answers

8140 [A] (B01) AMT-P
8141 [B] (B02) AMT-P
8142 [B] (B02) AMT-P
8143 [B] (B02) AMT-P
8144 [C] (B02) AMT-P
8145 [B] (B02) AMT-P
8146 [A] (B02) AMT-P
8147 [C] (B02) AMT-P
8148 [A] (B02) AMT-P

The can-type combustion chamber of a turbojet engine consists of an outer case, or housing and inside it is a perforated stainless steel combustion chamber or liner.

The case and liner are removed as a unit for routine maintenance.

8149. The diffuser section of a jet engine is located between

A—the burner section and the turbine section.
B—station No. 7 and station No. 8.
C—the compressor section and the burner section.

The diffuser is an annular chamber fitted with a number of vanes that form a series of divergent passages between the centrifugal compressor and the burner section.

As the diffuser vanes direct the flow of air into the burners, they increase the pressure of the air and decrease its velocity.

8150. When the leading edge of a first-stage turbine blade is found to have stress rupture cracks, which of the following should be suspected?

A—Faulty cooling shield.
B—Overtemperature condition.
C—Overspeed condition.

Stress rupture cracks or deformation of the leading edge of the first-stage turbine blades are usually caused by an overtemperature condition.

Overtemperature operation must be suspected when finding this type of damage.

8151. Turbine blades are generally more susceptible to operating damage than compressor blades because of

A—higher centrifugal loading.
B—exposure to high temperatures.
C—high pressure and high velocity gas flow.

Turbine blades are generally more susceptible to operating damage than compressor blades because they are continually exposed to such extremely high temperatures.

8152. Which of the following is the ultimate limiting factor of turbine engine operation?

A—Compressor inlet air temperature.
B—Turbine inlet temperature.
C—Burner-can pressure.

The turbine-inlet temperature (TIT) is the ultimate limiting factor in the operation of a turbojet engine. The temperature in a turbine engine is the highest at the inlet of the turbine.

8153. The recurrent ingestion of dust or other fine airborne particulates into a turbine engine can result in

A—foreign object damage to the compressor section.
B—the need for less frequent abrasive grit cleaning of the engine.
C—erosion damage to the compressor and turbine sections.

Dust and other fine airborne particulates can damage a turbine engine by eroding components in the compressor and turbine stages.

Turbine engines operated in a dusty environment normally have efficient dust and particle separators in their induction sections.

8154. Which of the following engine variables is the most critical during turbine engine operation?

A—Compressor inlet air temperature.
B—Compressor RPM.
C—Turbine inlet temperature.

Turbine-inlet temperature is the highest temperature inside a turbine engine. Therefore, it is the most critical variable of engine operation.

It is impractical to measure turbine-inlet temperature in most engines. So, the temperature-measuring thermocouples are usually installed at the turbine discharge. The turbine-outlet temperature gives a relative indication of the temperature at the turbine inlet.

Therefore, if the turbine-outlet temperatures are kept within range, it can be assumed that the turbine-inlet temperatures are also within range.

8155. Reduced blade vibration and improved airflow characteristics in gas turbines are brought about by

A—fir-tree blade attachment.
B—impulse type blades.
C—shrouded turbine rotor blades.

Shrouded turbine blades are used to reduce blade vibration and improve the airflow characteristics through the turbine.

8156. Which turbine engine compressor offers the greatest advantages for both starting flexibility and improved high-altitude performance?

A—Dual-stage, centrifugal-flow.
B—Split-spool, axial-flow.
C—Single-spool, axial-flow.

The two-spool (split-spool) axial-flow compressor offers the greatest starting flexibility and improved high-altitude performance of any of the gas turbine engine configurations.

Answers

8149 [C] (B02) AMT-P	8150 [B] (B02) AMT-P	8151 [B] (B02) AMT-P	8152 [B] (B02) AMT-P
8153 [C] (B02) AMT-P	8154 [C] (B02) AMT-P	8155 [C] (B02) AMT-P	8156 [B] (B02) AMT-P

8157. Jet engine turbine blades removed for detailed inspection must be reinstalled in

A—a specified slot 180° away.
B—a specified slot 90° away in the direction of rotation.
C—the same slot.

Turbine blades are individually weighed and coded for installation in the disks in such a way that they best distribute the weight evenly around the disk.

If a blade is removed from a disk for inspection, it must be reinstalled in the same slot from which it was removed.

8158. An advantage of the centrifugal-flow compressor is its high

A—pressure rise per stage.
B—ram efficiency.
C—peak efficiency.

A centrifugal compressor is simple and rugged and it can be made at a relatively low cost.

The pressure rise, which is produced by expansion of the gas in the diffuser manifold and by the conversion of kinetic energy of motion into static pressure, is high for each stage.

8159. The highest heat-to-metal contact in a jet engine is the

A—burner cans.
B—turbine inlet guide vanes.
C—turbine blades.

The highest heat-to-metal contact inside a jet engine occurs at the entrance to the first stage of the turbine. This is at the turbine-inlet guide vanes.

8160. Which two elements make up the axial-flow compressor assembly?

A—Rotor and stator.
B—Compressor and manifold.
C—Stator and diffuser.

An axial-flow compressor is made up of rotors (the rotating part of the compressor) and stators (the stationary part of the compressor).

8161. The two types of centrifugal compressor impellers are

A—single entry and double entry.
B—rotor and stator.
C—impeller and diffuser.

The two types of centrifugal compressors used in turbojet engines are single-entry and double-entry.

A single-entry compressor has scrolls on only one side, while a double-entry compressor has scrolls on both sides.

A double-entry compressor is much like two single-entry compressors back to back.

8162. Between each row of rotating blades in a turbine engine compressor, there is a row of stationary blades which act to diffuse the air. These stationary blades are called

A—buckets.
B—rotors.
C—stators.

The stationary blades between each set of rotating blades in an axial-flow turbine-engine compressor are called stators.

The function of the stators is to receive the air from each stage of the compressor and deliver it to the next stage at the proper velocity, direction, and pressure.

8163. Standard sea level pressure is

A—29.00" Hg.
B—29.29" Hg.
C—29.92" Hg.

Standard sea-level atmospheric pressure is 29.92 inches of mercury ("Hg), 760 millimeters of mercury (mm Hg), 1013.2 millibars (mb), or 14.69 pounds per square inch (psi).

8164. Using standard atmospheric conditions, the standard sea level temperature is

A—59°F.
B—59°C.
C—29°C.

The standard sea-level temperature for computing the power of a gas turbine engine is 15° Celsius, or 59° Fahrenheit.

8165. When aircraft turbine blades are subjected to excessive heat stress, what type of failures would you expect?

A—Bending and torsion.
B—Torsion and tension.
C—Stress rupture.

A turbine that has been subjected to excessive temperatures is likely to have blades that show indications of stress-rupture failure.

Answers

8157 [C] (B02) AMT-P	8158 [A] (B02) AMT-P	8159 [B] (B02) AMT-P	8160 [A] (B02) AMT-P
8161 [A] (B02) AMT-P	8162 [C] (B02) AMT-P	8163 [C] (B02) AMT-G	8164 [A] (B02) AMT-G
			8165 [C] (B02) AMT-P

Stress-rupture cracks appear as minute hairline cracks on or across the leading edge of the blade at right angles to the edge.

8166. In an axial-flow compressor, one purpose of the stator vanes at the discharge end of the compressor is to

A—straighten the airflow and eliminate turbulence.
B—increase the velocity and prevent swirling and eddying.
C—decrease the velocity, prevent swirling, and decrease pressure.

The stator vanes located at the discharge end of an axial-flow compressor are used to straighten the airflow and eliminate turbulence of the air as it enters the combustors.

8167. Compressor field cleaning on turbine engines is performed primarily in order to

A—prevent engine oil contamination and subsequent engine bearing wear or damage.
B—facilitate flight line inspection of engine inlet and compressor areas for defects or FOD.
C—prevent engine performance degradation, increased fuel costs, and damage or corrosion to gas path surfaces.

Foreign deposits on the compressor rotor and stator vanes reduce aerodynamic efficiency of the blades, degrade the engine performance, and increase fuel costs. Compressor field cleaning removes salt or dirt deposits from the blades and vanes, restores efficiency, and prevents corrosion of the surfaces along the gas path.

8168. Hot section inspections for many modern turbine engines are required

A—only at engine overhaul.
B—only when an overtemperature or overspeed has occurred.
C—on a time or cycle basis.

A hot-section inspection of a gas turbine engine consists of a visual and dimensional inspection of components in the hot section, which includes the combustion section, turbine inlet guide vanes, turbine wheels, and all of the other components that operate in the high-temperature gas path.

Hot-section inspections are regularly conducted on a time or cycle basis, and when there has been a deterioration of certain of the engine parameters, such as EGT, EPR, or fuel flow.

8169. A purpose of the shrouds on the turbine blades of an axial-flow engine is to

A—reduce vibration.
B—increase tip speed.
C—reduce air entrance.

Some turbine blades have shrouds on their outer ends that contact the shroud on the adjacent blade. These shrouds form a band around the outer perimeter of the wheel that improves the efficiency of the wheel and reduces the vibration of the blades.

8170. In a dual axial-flow compressor, the first stage turbine drives

A—N_2 compressor.
B—N_1 compressor.
C—low pressure compressor.

The first-stage turbine in a twin-spool turbojet engine drives the high-pressure compressor, the N_2 compressor.

8171. What should be done initially if a turbine engine catches fire when starting?

A—Turn off the fuel and continue engine rotation with the starter.
B—Continue engine start rotation and discharge a fire extinguisher into the intake.
C—Continue starting attempt in order to blow out the fire.

If a turbine engine catches fire in the process of starting, turn the fuel off and continue rotating the engine with the starter to force enough air through the engine to blow the fire out.

8172. What is the proper starting sequence for a turbojet engine?

A—Ignition, starter, fuel.
B—Starter, ignition, fuel.
C—Starter, fuel, ignition.

The proper sequence for starting a turbojet engine is to engage the starter to start the compressor turning, then turn on the ignition and finally, turn on the fuel.

8173. A weak fuel to air mixture along with normal airflow through a turbine engine may result in

A—a rich flameout.
B—a lean die-out.
C—high EGT.

If the fuel furnished to a turbine engine is decreased to the point that there is a weak fuel to air mixture with a normal flow of air through the engine, there is a danger of a lean die-out.

Answers

8166 [A] (B02) AMT-P	8167 [C] (B02) AMT-P	8168 [C] (B02) AMT-P	8169 [A] (B02) AMT-P
8170 [A] (B02) AMT-P	8171 [A] (B02) AMT-P	8172 [B] (B02) AMT-P	8173 [B] (B02) AMT-P

8174. What is used in turbine engines to aid in stabilization of compressor airflow during low thrust engine operation?

A—Stator vanes and rotor vanes.
B—Variable guide vanes and/or compressor bleed valves.
C—Pressurization and dump valves.

Some axial-flow gas turbine engines use variable inlet guide vanes and bleed-air valves to stabilize the airflow through the compressor during low-thrust operations.

If the compressor RPM is high, relative to the amount of air flowing through the engine, the angle of attack of the compressor blades will become excessive and a compressor surge or stall can develop.

To prevent a compressor stall, the inlet guide vanes are turned to the correct angle, and the bleed-air valves are automatically opened by actuators controlled by the fuel control. The bleed-air valves reduce the back pressure and allow more air to flow through the compressor to reduce the angle of attack of the compressor blades.

8175. In a turbine engine with a dual-spool compressor, the low speed compressor

A—always turns at the same speed as the high speed compressor.
B—is connected directly to the high speed compressor.
C—seeks its own best operating speed.

The high-pressure rotor of a two-spool (split-spool) compressor is governed for speed, but the low-pressure rotor is free to operate at its own best speed.

As the air density decreases at altitude, the compressor load decreases and the low-pressure rotor increases its speed.

8176. What is the function of the inlet guide vane assembly on an axial-flow compressor?

A—Directs the air into the first stage rotor blades at the proper angle.
B—Converts velocity energy into pressure energy.
C—Converts pressure energy into velocity energy.

Inlet guide vanes used with an axial-flow compressor change the angle of airflow to direct it into the first stage rotor blades at the proper angle. The inlet guide vanes do not change either the velocity or the pressure of the inlet air.

8177. Hot spots on the tail cone of a turbine engine are possible indicators of a malfunctioning fuel nozzle or

A—a faulty combustion chamber.
B—a faulty igniter plug.
C—an improperly positioned tail cone.

Hot spots, which are localized areas of overheating in the tail cone of a gas turbine engine, are usually caused by a malfunctioning fuel nozzle or a faulty combustion chamber that prevent a uniform flow of cooling air.

8178. The stator vanes in an axial-flow compressor

A—convert velocity energy into pressure energy.
B—convert pressure energy into velocity energy.
C—direct air into the first stage rotor vanes at the proper angle.

Stator vanes in an axial-flow compressor convert velocity energy that has been put into the air by the rotors into pressure energy.

The stator vanes also direct the air into the following stage of rotor blades in the correct direction.

8179. The velocity of subsonic air as it flows through a convergent nozzle

A—increases.
B—decreases.
C—remains constant.

When subsonic air flows through a convergent nozzle, its velocity increases and its pressure decreases.

This action is in accordance with Bernoulli's principle, which gives us the relationship between kinetic energy and potential energy in a column of moving fluid. The kinetic energy relates to the velocity of the fluid and the potential energy relates to the pressure of the fluid.

8180. The velocity of supersonic air as it flows through a divergent nozzle

A—increases.
B—decreases.
C—is inversely proportional to the temperature.

When supersonic air flows through a divergent nozzle (a nozzle whose cross-sectional area increases in the direction of air flow), its velocity increases and its pressure decreases.

Answers

8174 [B] (B02) AMT-P	8175 [C] (B02) AMT-P	8176 [A] (B02) AMT-P	8177 [A] (B02) AMT-P
8178 [A] (B02) AMT-P	8179 [A] (B02) AMT-STRUC	8180 [A] (B02) AMT-STRUC	

8181. The pressure of subsonic air as it flows through a convergent nozzle

A— increases.
B— decreases.
C— remains constant.

When subsonic air flows through a convergent nozzle, its velocity increases and its pressure decreases.

This action is in accordance with Bernoulli's principle, which gives us the relationship between kinetic energy and potential energy in a column of moving fluid. The kinetic energy relates to the velocity of the fluid and the potential energy relates to the pressure of the fluid.

8182. The pressure of supersonic air as it flows through a divergent nozzle

A— increases.
B— decreases.
C— is inversely proportional to the temperature.

When supersonic air flows through a divergent nozzle (a nozzle whose cross-sectional area increases in the direction of air flow), its velocity increases and its pressure decreases.

8183. Anti-icing of jet engine air inlets is commonly accomplished by

A— electrical heating elements inside the inlet guide vanes.
B— engine bleed air ducted through the critical areas.
C— electrical heating elements located within the engine alr Inlet cowling.

Anti-icing (preventing the formation of ice) is accomplished on the air inlets of a turbojet engine by ducting hot engine bleed-air through the areas on which the ice is likely to form.

8184. Generally, when starting a turbine engine, the starter should be disengaged

A— after the engine has reached self-accelerating speed.
B— only after the engine has reached full idle RPM.
C— when the ignition and fuel system are activated.

When starting a turbojet engine, the starter is left engaged until the engine reaches its self-accelerating speed.

If the starter is disengaged too soon, the engine may fail to accelerate to its idle RPM, and a "hung start" results.

8185. What is the primary advantage of an axial-flow compressor over a centrifugal compressor?

A— High frontal area.
B— Less expensive.
C— Greater pressure ratio.

The primary advantage of an axial-flow compressor over a centrifugal compressor is that axial-flow compressors are capable of producing higher pressure ratios and they have relatively high efficiencies.

8186. The purpose of a bleed valve, located in the beginning stages of the compressor, in an aircraft gas turbine engine is to

A— vent some of the air overboard to prevent a compressor stall.
B— control excessively high RPM to prevent a compressor stall.
C— vent high ram air pressure overboard to prevent a compressor stall.

Some aircraft gas turbine engines have automatic bleed valves that prevent compressor stall or surge by venting overboard some of the air from the compressor. This prevents the air from "piling up" in the higher-pressure stages of the compressor and restricting the flow of air through the engine.

8187. What is meant by a double entry centrifugal compressor?

A— A compressor that has two intakes.
B— A two-stage compressor independently connected to the main shaft.
C— A compressor with vanes on both sides of the impeller.

A double-entry centrifugal compressor is a compressor that has vanes on both sides of the impeller.

The intake air is ducted into the impeller at both its front and back sides. The accelerated air is taken from its rim.

8188. What is the major function of the turbine assembly in a turbojet engine?

A— Directs the gases in the proper direction to the tailpipe.
B— Supplies the power to turn the compressor.
C— Increases the temperature of the exhaust gases.

The major function of the turbine assembly in a turbojet engine is that of supplying power to turn the compressor.

Answers

8181 [B] (B02) AMT-STRUC
8182 [B] (B02) AMT-STRUC
8183 [B] (B02) AMT-P
8184 [A] (B02) AMT-P
8185 [C] (B02) AMT-P
8186 [A] (B02) AMT-P
8187 [C] (B02) AMT-P
8188 [B] (B02) AMT-P

8189. Stator blades in the compressor section of an axial-flow turbine engine

A—increase the air velocity and prevent swirling.
B—straighten the airflow and accelerate it.
C—decrease the air velocity and prevent swirling.

Stator vanes in an axial-flow compressor convert velocity energy that has been put into the air by the rotors into pressure energy. They slow the air, which increases its pressure.

The stator vanes also direct the air into the following stage of rotor blades in the correct direction.

8190. A gas turbine engine comprises which three main sections?

A—Compressor, diffuser, and stator.
B—Turbine, combustion, and stator.
C—Turbine, compressor, and combustion.

The three main sections of a turbine engine are: the compressor, which increases the pressure of the air entering the engine; the combustors, in which energy from burning fuel is added to the compressed air; and the turbine, which drives the compressor.

8191. What type of turbine blade is most commonly used in aircraft jet engines?

A—Reaction.
B—Impulse.
C—Impulse-reaction.

The most common type of turbine blades used in modern aircraft jet engines is the impulse-reaction type.

An impulse-reaction turbine blade has an impulse section at its root and a reaction section at its tip.

The exit pressure of a impulse-reaction turbine blade is relatively constant across its length.

8192. What is the primary factor which controls the pressure ratio of an axial-flow compressor?

A—Number of stages in compressor.
B—Compressor inlet pressure.
C—Compressor inlet temperature.

The pressure ratio of an axial-flow compressor is a function of the number of stages of compression (the number of stages of compressor rotors and stators).

8193. The non-rotating axial-flow compressor airfoils in an aircraft gas turbine engine, are called

A—pressurization vanes.
B—stator vanes.
C—bleed vanes.

The nonrotating airfoils in an aircraft gas turbine engine axial-flow compressor are called stator vanes.

8194. (1) In a turbine engine axial-flow compressor, each consecutive pair of rotor and stator blades constitutes a pressure stage.

(2) In a turbine engine axial-flow compressor, the number of rows of stages is determined by the amount of air and total pressure rise required.

Regarding the above statements,

A—only No. 1 is true.
B—only No. 2 is true.
C—both No. 1 and No. 2 are true.

Both statements are true. Each consecutive pair of rotor blades and stator vanes in an axial-flow compressor constitutes a pressure stage.

The number of pressure stages in the compressor is determined by the amount of air and the total pressure rise required by the engine.

8195. The air passing through the combustion chamber of a turbine engine is

A—used to support combustion and to cool the engine.
B—entirely combined with fuel and burned.
C—speeded up and heated by the action of the turbines.

Part of the air flowing through the combustion chamber of a jet engine mixes with the fuel for combustion. The majority of the air, however, passes between the outer casing and the liner and is used to cool the combustion gases.

8196. The stators in the turbine section of a gas turbine engine

A—increase the velocity of the gas flow.
B—decrease the velocity of the gas flow.
C—increase the pressure of the gas flow.

The turbine in a gas turbine engine extracts energy from the burning gases as they pass through it. The stators in front of the rotating turbine wheels increase the velocity of the gases and direct them so they will strike the rotors at the correct angle.

Answers

8189 [C] (B02) AMT-P	8190 [C] (B02) AMT-P	8191 [C] (B02) AMT-P	8192 [A] (B02) AMT-P
8193 [B] (B02) AMT-P	8194 [C] (B02) AMT-P	8195 [A] (B02) AMT-P	8196 [A] (B02) AMT-P

8197. The compressor stators in a gas turbine engine act as diffusers to

A—decrease the velocity of the gas flow.
B—increase the velocity of the gas flow.
C—increase the velocity and decrease the pressure of the gas.

The stator vanes in an axial-flow compressor are placed to the rear of the rotor blades to receive the high-velocity air and act as diffusers. They change some of the kinetic energy of velocity into potential energy of pressure.

8198. The procedure for removing the accumulation of dirt deposits on compressor blades is called

A—the soak method.
B—field cleaning.
C—the purging process.

Field cleaning of a turbine engine is the removal of contaminants from the blades of the compressor by running wash water or an abrasive through the engine.

8199. Which of the following may be used to accomplish internal inspection of an assembled turbine engine?

1. Infrared photography.
2. Ultrasound.
3. A borescope.
4. Fluorescent penetrant and ultraviolet light.

A—1, 2, 3.
B—1, 3.
C—3.

The inspection of the interior of a turbine engine installed on an aircraft can be performed by the use of a borescope.

8200. What is the possible cause when a turbine engine indicates no change in power setting parameters, but oil temperature is high?

A—High scavenge pump oil flow.
B—Engine main bearing distress.
C—Turbine damage and/or loss of turbine efficiency.

Too high a lubricating-oil temperature, with no change in the power setting parameters, could indicate that the main bearing is overheating and this heat is being absorbed in the oil.

8201. Newton's First Law of Motion, generally termed the Law of Inertia, states:

A—To every action there is an equal and opposite reaction.
B—Force is proportional to the product of mass and acceleration.
C—Every body persists in its state of rest, or of motion in a straight line, unless acted upon by some outside force.

Newton's First Law of Motion tells us that every body at rest will try to remain at rest and every body in motion will try to remain in motion in a straight line, at the same speed, unless it is acted upon by an outside force.

8202. A turbine engine hot section is particularly susceptible to which kind of damage?

A—Scoring.
B—Cracking.
C—Galling.

Cracking is one of the most widely found forms of damage in the hot section of a gas turbine engine.

Vibration and the extremes of temperature cause the thin metal of which hot-section components are made, to crack.

8203. Dirt particles in the air being introduced into the compressor of a turbine engine will form a coating on all but which of the following?

A—Turbine blades.
B—Casings.
C—Inlet guide vanes.

Dirt particles introduced into the turbine engine will cause a coating to form on the casings, inlet guide vanes and compressor blades.

The extreme heat in the turbine section prevents the coating from forming on the turbine blades.

8204. Severe rubbing of turbine engine compressor blades will usually cause

A—bowing.
B—cracking.
C—galling.

Severe rubbing of the turbine-engine compressor blades will usually result in galling, which is a transfer of metal from one surface to another.

Answers

8197 [A] (B02) AMT-P	8198 [B] (B02) AMT-P	8199 [C] (B02) AMT-P	8200 [B] (B03) AMT-P
8201 [C] (B03) AMT-G	8202 [B] (B03) AMT-P	8203 [A] (B03) AMT-P	8204 [C] (B03) AMT-P

8205. Which of the following influences the operation of an automatic fuel control unit on a turbojet engine?

A—Burner pressure.
B—Mixture control position.
C—Exhaust gas temperature.

Most automatic fuel control units for turbojet engines sense inlet air temperature, compressor RPM, burner pressure (compressor discharge pressure) and the position of the power lever (throttle).

8206. If a turbine engine is unable to reach takeoff EPR before its EGT limit is reached, this is an indication that the

A—fuel control must be replaced.
B—EGT controller is out of adjustment.
C—compressor may be contaminated or damaged.

When a turbine engine compressor is contaminated or damaged, the airflow is disturbed, and the EGT limit may be reached before takeoff EPR is attained.

This condition may be corrected by field cleaning the compressor or retrimming the fuel control.

8207. The Brayton cycle is known as the constant

A—pressure cycle.
B—temperature cycle.
C—mass cycle.

The Brayton cycle of energy release used in a gas turbine engine is known as the constant-pressure cycle.

Energy added to the air flowing through the engine by the burning fuel, causes the volume of the air to increase, but the pressure remains relatively constant.

8208. Continued and/or excessive heat and centrifugal force on turbine engine rotor blades is likely to cause

A—profile.
B—creep.
C—galling.

Creep, which is a permanent elongation of the turbine blades caused by heat loads and centrifugal loads, is likely to occur when the engine has been exposed to continued and/or excessive heat.

8209. If the RPM of an axial-flow compressor remains constant, the angle of attack of the rotor blades can be changed by

A—changing the velocity of the airflow.
B—changing the compressor diameter.
C—increasing the pressure ratio.

Two factors affecting the angle of attack of an axial-flow compressor blade are the velocity of the air through the engine and the RPM of the compressor.

If the airflow entering the engine is restricted, reducing its velocity, the angle of attack of the compressor blade will increase to such a point that compressor stall can occur.

8210. The compression ratio of an axial-flow compressor is a function of the

A—number of compressor stages.
B—rotor diameter.
C—air inlet velocity.

The compression ratio of an axial-flow compressor is determined by the number of stages of compression.

8211. Which of the following variables affect the inlet air density of a turbine engine?

1. Speed of the aircraft.
2. Compression ratio.
3. Turbine inlet temperature.
4. Altitude of the aircraft.
5. Ambient temperature.
6. Turbine and compressor efficiency.

A—1, 3, 6.
B—1, 4, 5.
C—4, 5, 6.

Three factors that affect the density of the air taken into the inlet air system of a turbojet engine are the speed of the aircraft, the altitude at which the aircraft is flying and the ambient (surrounding) air temperature.

8212. Which of the following factors affect the thermal efficiency of a turbine engine?

1. Turbine inlet temperature.
2. Compression ratio.
3. Ambient temperature.
4. Speed of the aircraft.
5. Turbine and compressor efficiency.
6. Altitude of the aircraft.

A—3, 4, 6.
B—1, 2, 5.
C—1, 2, 6.

Answers

8205 [A] (B03) AMT-P	8206 [C] (B03) AMT-P	8207 [A] (B03) AMT-P	8208 [B] (B03) AMT-P
8209 [A] (B03) AMT-P	8210 [A] (B03) AMT-P	8211 [B] (B03) AMT-P	8212 [B] (B03) AMT-P

Three factors that affect the thermal efficiency of a turbine engine are the turbine-inlet temperature, the compression ratio of the compressor and the turbine and compressor efficiency.

8213. Why do some turbine engines have more than one turbine wheel attached to a single shaft?

A—To facilitate balancing of the turbine assembly.
B—To help stabilize the pressure between the compressor and the turbine.
C—To extract more power from the exhaust gases than a single wheel can absorb.

Some turbine engines have more than one turbine wheel on a single shaft in order to extract more power from the exhaust gases than a single turbine wheel can absorb.

8214. The exhaust section of a turbine engine is designed to

A—impart a high exit velocity to the exhaust gases.
B—increase temperature, therefore increasing velocity.
C—decrease temperature, therefore decreasing pressure.

The exhaust section of a turbojet engine is designed in such a way that it gives a high velocity to the exhaust gases leaving the engine.

8215. Which of the following types of combustion sections are used in aircraft turbine engines?

A—Annular, variable, and cascade vane.
B—Can, multiple-can, and variable.
C—Multiple-can, annular, and can-annular.

Three basic types of combustion sections used in gas turbine engines are the multiple-can type, the annular type, and the can-annular type.

8216. A cool-off period prior to shutdown of a turbine engine is accomplished in order to

A—allow the turbine wheel to cool before the case contracts around it.
B—prevent vapor lock in the fuel control and/or fuel lines.
C—prevent seizure of the engine bearings.

A rule of thumb for turbine engine operation: when an engine has been operated above approximately 85% RPM for periods longer than one minute; during the last five minutes before shutdown, the engine should be operated below 85% RPM (preferably at idle) for a period of five minutes. This prevents the possibility of the engine case contracting around the turbine wheels before they have cooled and contracted to their normal size.

It is also important that all surfaces contacted with engine oil be cooled to their normal operating temperature, to prevent oil left on a hot surface from coking (turning into a hard carbon deposit).

8217. What type igniter plug is used in the low tension ignition system of an aircraft turbofan engine?

A—Low voltage, high amperage glow plug.
B—Self-ionizing or shunted-gap type plug.
C—Recessed surface gap plug.

Self-ionizing shunted-gap igniters are used in the low-tension ignition system of some aircraft turbofan engines. These igniters have a ceramic semiconductor material between the center electrode and the shell. The resistance of this semiconductor is low when it is relatively cool, but it increases as it heats up. When the storage capacitor discharges through the igniter, the current initially flows to ground through the semiconductor, which gets so hot it becomes incandescent and its resistance increases. The air gap between the electrodes becomes ionized, and its resistance drops below that of the semiconductor. The remainder of the current discharges across the air gap in a surge as a high-energy spark.

8218. What is meant by a shrouded turbine?

A—The turbine blades are shaped so that their ends form a band or shroud.
B—The turbine wheel is enclosed by a protective shroud to contain the blades in case of failure.
C—The turbine wheel has a shroud or duct which provides cooling air to the turbine blades.

A shrouded turbine is one in which each of the blades is made in the shape of the letter T. Each bar on the end of the blades touches the other to form a band, or shroud, around the turbine wheel.

The shrouds increase the efficiency of the turbine and improve the vibration characteristics of the turbine blades.

8219. What term is used to describe a permanent and cumulative deformation of the turbine blades of a turbojet engine?

A—Stretch.
B—Distortion.
C—Creep.

Creep is the stretching (elongation) of a turbine blade caused by prolonged exposure to high temperatures and centrifugal force.

Answers

8213 [C] (B03) AMT-P	8214 [A] (B03) AMT-P	8215 [C] (B03) AMT-P	8216 [A] (B03) AMT-P
8217 [B] (B03) AMT-P	8218 [A] (B03) AMT-P	8219 [C] (B03) AMT-P	

8220. What is the purpose of the dump valve used on aircraft gas turbine engines?

A—The fuel is quickly cut off to the nozzles and the manifolds are drained preventing fuel boiling off as a result of residual engine heat.
B—The valve controls compressor stall by dumping compressor bleed air from the compressor discharge port under certain conditions.
C—Maintains minimum fuel pressure to the engine fuel control unit inlet and dumps excessive fuel back to the inlet of the engine-driven fuel pump.

The dump valve, which is a portion of the pressurizing and dump valve assembly, dumps the fuel from the fuel manifold when the engine is shut down. Dumping this fuel sharply cuts off combustion and prevents the fuel boiling as a result of residual engine heat.

8221. At what stage in a turbine engine are gas pressures the greatest?

A—Compressor inlet.
B—Turbine outlet.
C—Compressor outlet.

The gas pressure inside a turbine engine is the greatest at the compressor outlet.

8222. In what section of a turbojet engine is the jet nozzle located?

A—Combustion.
B—Turbine.
C—Exhaust.

The rear opening of a turbine-engine exhaust duct is called the exhaust nozzle (jet nozzle).

The nozzle acts as an orifice, the size of which determines the density and velocity of the gases as they leave the engine.

8223. (1) Accumulation of contaminates in the compressor of a turbojet engine reduces aerodynamic efficiency of the blades.

(2) Two common methods for removing dirt deposits from turbojet engine compressor blades are a fluid wash and an abrasive grit blast.

Regarding the above statements,

A—only No. 1 is true.
B—only No. 2 is true.
C—both No. 1 and No. 2 are true.

Both statements are true.

Accumulation of contaminants in the compressor reduces aerodynamic efficiency of the blades and reduces engine performance.

Two common methods for removing dirt deposits are a fluid wash and an abrasive grit blast.

8224. Hot spots in the combustion section of a turbojet engine are possible indicators of

A—faulty igniter plugs.
B—dirty compressor blades.
C—malfunctioning fuel nozzles.

Hot spots are possible indicators of a serious condition, such as malfunctioning fuel nozzles or other fuel-system malfunctions.

8225. Which of the following can cause fan blade shingling in a turbofan engine?

1. Engine overspeed.
2. Engine overtemperature.
3. Large, rapid throttle movements.
4. FOD.

A—1, 2.
B—1, 2, 3, 4.
C—1, 4.

Fan blade shingling is a condition of the fan in a turbofan engine when the midspan shrouds on the fan blades overlap in much the same way shingles on a roof overlap.

Fan blade shingling is caused by the rotating fan encountering opposition such as engine stall, bird strike, foreign object damage (FOD), or by engine overspeed.

8226. Compressor stall is caused by

A—a low angle of attack airflow through the first stages of compression.
B—a high angle of attack airflow through the first stages of compression.
C—rapid engine deceleration.

A compressor stalls when the angle of attack of the blades becomes excessive.

Two factors affecting the angle of attack of an axial-flow compressor blade are the velocity of the air through the engine and the RPM of the compressor.

If the airflow entering the engine is restricted, reducing its velocity, the angle of attack of the compressor blade will increase to such a point that compressor stall can occur.

Answers

8220 [A] (B03) AMT-P	8221 [C] (B03) AMT-P	8222 [C] (B03) AMT-P	8223 [C] (B03) AMT-P
8224 [C] (B03) AMT-P	8225 [C] (B03) AMT-P	8226 [B] (B03) AMT-P	

8227. A condition known as "hot streaking" in turbine engines is caused by

A—a partially clogged fuel nozzle.
B—a misaligned combustion liner.
C—excessive fuel flow.

Hot streaking is a hot-section condition in which the flame penetrates through the entire turbine system to the tail pipe.

Hot streaking is caused by a partially clogged fuel nozzle which does not atomize the fuel into a cone-shaped pattern but rather, allows a small fuel stream to flow with sufficient force to cut through the cooling air blanket and impinge directly on the turbine surfaces.

8228. (Refer to Figure 1.) Determine which portion of the AD is applicable for Model O-690 series engine, serial No. 5863-40 with 283 hours' time in service.

A—(B), (1).
B—(A).
C—(B), (2).

This question requires some close examination.

Statement (A) does not apply because of the serial number of the engine. The serial number of the engine we are concerned with is 5863-40, and in statement (A), the engines have serial numbers from 101-40 through 5264-40.

Statement (B) does apply to this engine, because the serial number falls within the range of numbers given.

(1) is applicable, because the engine has more than 275 hours in service. It has 283 hours in service.

(2) does not apply, because this engine has more than 275 hours time in service.

8229. A Cessna 180 aircraft has a McCauley propeller Model No. 2A34C50/90A. The propeller is severely damaged in a ground accident, and this model propeller is not available for replacement. Which of the following should be used to find an approved alternate replacement?

A—Summary of Supplemental Type Certificates.
B—Aircraft Specifications/Type Certificate Data Sheets.
C—Aircraft Engine and Propeller Specifications/Type Certificate Data Sheets.

The Aircraft Specifications or Type Certificate Data Sheets list all of the propellers that are approved for a specific airplane.

This is the compliance portion of an FAA Airworthiness Directive.

Compliance required as indicated:

(A) For model O-690 series engines, serial Nos. 101-40 through 5264-40 and IO-690 series engines, serial Nos. 101-48 through 423-48, compliance with (C) required within 25 hours' time in service after the effective date of this AD and every 100 hours' time in service thereafter.

(B) For Model O-690 series engines, serial Nos. 5265-40 through 6129-40 and IO-690 series engines, serial Nos. 424-48 through 551-48, compliance with (C) required as follows:

(1) Within 25 hours' time in service after the effective date of this AD and every 100 hours' time in service thereafter for engines with more than 275 hours' time in service on the effective date of this AD.

(2) Prior to the accumulation of 300 hours total time in service and every 100 hours' time in service thereafter for engines with 275 hours or less time in service on the effective date of this AD.

(C) Inspect the oil pump drive shaft (P/N 67512) on applicable engines in accordance with instructions contained in Connin Service Bulletin No. 295. Any shafts which are found to be damaged shall be replaced before further flight. These inspections shall be continued until Connin P/N 67512 (redesigned) or P/N 74641 oil pump drive shaft is installed at which time the inspections may be discontinued.

Figure 1

Answers
8227 [A] (B03) AMT-P
8228 [A] (C01) AC 39-7B
8229 [B] (C01) AMT-G

8230. Which of the following is used to monitor the mechanical integrity of the turbines, as well as to check engine operating conditions of a turbine engine?

A—Engine oil pressure.
B—Exhaust gas temperature.
C—Engine pressure ratio.

Temperature is one of the most important considerations in the operation of a turbine engine.

Because of this, the exhaust-gas temperature, which gives us an indication of the turbine-inlet temperature, allows us to monitor the mechanical integrity of the turbines as well as to check the operating conditions of the engine.

8231. On a reciprocating engine aircraft using a shrouded exhaust muffler system as a source for cabin heat, the exhaust system should be

A—visually inspected for any indication of cracks or an operational carbon monoxide detection test should be done.
B—replaced at each reciprocating engine overhaul by a new or overhauled exhaust system or an hydrostatic test should be accomplished.
C—removed and the exhaust muffler checked for cracks by using magnetic particle inspection method or an hydrostatic test should be done on the exhaust muffler.

Exhaust systems that provide heat for the aircraft cabin should periodically be visually inspected for cracks with the shrouds removed. Carbon monoxide detectors should be installed in the cabins of all aircraft using this type of heat.

8232. (1) Airworthiness Directives are Federal Aviation Regulations and must be complied with unless specific exemption is granted.

(2) Airworthiness Directives of an emergency nature may require immediate compliance upon receipt.

Regarding the above statements,

A—only No. 1 is true.
B—only No. 2 is true.
C—both No. 1 and No. 2 are true.

Statement (1) is true. An Airworthiness Directive (AD note) is issued by the Federal Aviation Administration anytime a condition arises that prevents a certificated aircraft, engine, propeller, or appliance from continuing to comply with its airworthiness certification.

All applicable AD notes must be complied with unless a special exemption is granted.

Statement (2) is also true. When an AD is issued as an emergency AD note, it must normally be complied with before further flight.

8233. Which of the following contains a minimum checklist for 100-hour inspections of engines?

A—14 CFR Part 33 Appendix A.
B—14 CFR Part 43 Appendix D.
C—Engine Specifications and Type Certificate Data Sheets.

Federal Aviation Regulations, Part 43, entitled "Maintenance, Preventive Maintenance, Rebuilding and Alteration," contains in Appendix D the scope and details of items be included in an annual and 100-hour inspection.

The items included in this list are the absolute minimum that should be inspected.

8234. When must an Airworthiness Directive (AD) be complied with after it becomes effective?

A—As specified in the AD.
B—During the next scheduled inspection.
C—At the next scheduled overhaul.

An Airworthiness Directive includes not only the action that must be taken, but also the time limit in which this action must be taken.

The time limit required for compliance is part of each Airworthiness Directive.

8235. Which of the following contains a table that lists the engines to which a given propeller is adaptable?

A—Aircraft Type Certificate Data Sheets.
B—Propeller Type Certificate Data Sheets.
C—Engine Type Certificate Data Sheets.

Propeller Type Certificate Data Sheets list the engines on which a given propeller can be used.

8236. Which of the following component inspections is to be accomplished on a 100-hour inspection?

A—Check internal timing of magneto.
B—Check cylinder compression.
C—Check valve timing.

Of the alternatives listed with this question, only the cylinder compression check is included in 14 CFR 43, Appendix D, as required to be included on a 100-hour or annual inspection.

Answers

8230 [B] (C01) AMT-P	8231 [A] (C01) AMT-P	8232 [C] (C01) AMT-G	8233 [B] (C01) 14 CFR 43
8234 [A] (C01) 14 CFR 39.3 & AC 39-7B	8235 [B] (C01) AMT-G	8236 [B] (C01) 14 CFR 43	

8237. You are performing a 100-hour inspection on an R985-22 aircraft engine. What does the "985" indicate?

A—The total piston displacement of the engine.
B—The pistons will pump a maximum of 985 cubic inches of air per crankshaft revolution.
C—The total piston displacement of one cylinder.

Aircraft engines are normally identified by a letter indicating the cylinder arrangement and a number indicating the total piston displacement of the engine.

In this case, an R-985-22 engine is a radial engine that has 985 cubic inches of piston displacement. Dash 22 is the model of this particular engine.

8238. Where would one find type design information for an R1830-92 engine certificated under the Civil Air Regulations (CAR) and installed on a DC-3?

A—The Aircraft Specifications and Type Certificate Data Sheet.
B—The Aircraft Engine Specifications.
C—The Aircraft Engine Type Certificate Handbook.

The DC-3 and its R-1830-92 engines were certificated before the Federal Aviation Administration came into being. They were certificated by the older Civil Aeronautics Administration.

The specifications for the R-1830-92 engines are included in the CAA Engine Specifications 5E4.

More recently certificated engines have their specifications in Engine Type Certificate Data Sheets.

8239. Straightening nitrided crankshafts is

A—recommended.
B—not recommended.
C—approved by the manufacturer.

Nitrided crankshafts have an extremely hard surface on the bearing journals. Any attempt to straighten them will likely result in a rupture of the nitrided surface.

Rupturing the hardened surface will cause an eventual failure of the crankshaft.

8240. The breaking loose of small pieces of metal from coated surfaces, usually caused by defective plating or excessive loads, is called

A—flaking.
B—chafing.
C—brinelling.

Bearing flaking causes small pieces of metal to break away from the hardened surface of the bearing race or roller.

This flaking may be caused by excessive bearing loads or defective plating.

8241. Each powerplant installed on an airplane with a Standard Airworthiness Certificate must have been

A—type certificated.
B—manufactured under the TSO system.
C—originally certificated for that aircraft.

Any engine to be installed on an aircraft having a Standard Airworthiness Certificate must be a type-certificated engine.

8242. A severe condition of chafing or fretting in which a transfer of metal from one part to another occurs is called

A—scoring.
B—burning.
C—galling.

Galling is a severe condition of chafing or fretting that occurs when two pieces of metal rub together and transfer metal from one piece to the other.

8243. Indentations on bearing races caused by high static loads are known as

A—fretting.
B—brinelling.
C—galling.

Brinelling is a condition that occurs in a bearing when the race is indented by the balls or rollers.

Brinelling is caused by high static loads being imposed on the bearings.

8244. When inspecting an aircraft reciprocating engine what document is used to determine if the proper magnetos are installed?

A—Instruction for continued airworthiness issued by the engine manufacturer.
B—Engine Manufacturer's Maintenance Manual.
C—Aircraft Engine Specifications or Type Certificate Data Sheets.

Aircraft Engine Specifications and Type Certificate Data Sheets list the specifications under which a particular aircraft engine was certified.

The mechanic would have to refer to whichever of these documents applies to a particular engine to determine whether or not the engine conforms to its original type design.

Answers

8237 [A] (C01) AMT-P	8238 [B] (C01) AMT-P	8239 [B] (C01) AC 43.13-1B	8240 [A] (C01) AMT-P
8241 [A] (C01) 14 CFR 23.903	8242 [C] (C01) AMT-P	8243 [B] (C01) AMT-P	8244 [C] (C01) AMT-G

8245. Which of the following can inspect and approve an engine major repair for return to service?

A—Certificated mechanic with airframe and powerplant ratings.
B—Certificated mechanic with a powerplant rating.
C—Certificated mechanic with inspection authorization.

An engine major repair must be inspected for conformity to approved data and approved for return to service by a certificated mechanic who holds an Inspection Authorization.

8246. What publication is used for guidance to determine whether a powerplant repair is major or minor?

A—Airworthiness Directives.
B—Federal Aviation Regulations, Part 43, appendix A.
C—Technical Standard Orders.

Federal Aviation Regulations, Part 43, entitled "Maintenance, Preventive Maintenance, Rebuilding and Alteration," lists, in Appendix A, a number of maintenance operations and categorizes them as to whether they are major or minor repairs, major or minor alterations, or preventive maintenance.

8247. The airworthiness standards for the issue of type certificates for small airplanes with nine or less passenger seats in the normal, utility, and acrobatic categories may be found in the

A—Federal Aviation Regulations, Part 23.
B—Supplemental Type Certificate.
C—Federal Aviation Regulations, Part 21.

The airworthiness standards for small airplanes (those having a maximum certificated takeoff weight of 12,500 pounds or less), and having nine or less passenger seats is found in 14 CFR Part 23. This was formerly called Federal Aviation Regulations, Part 23.

8248. Which of the following contains approved data for performing a major repair to an aircraft engine?

A—Engine Type Certificate Data Sheets.
B—Supplemental Type Certificates.
C—Manufacturer's maintenance instructions when FAA approved.

The engine maintenance instructions issued by the engine manufacturer and approved by the Federal Aviation Administration can be used as approved data when performing a major repair to an aircraft engine.

8249. What maintenance record(s) is/are required following a major repair of an aircraft engine?

A—Entries in engine maintenance records and a list of discrepancies for the FAA.
B—Entries in the engine maintenance record and FAA Form 337.
C—Entry in logbook.

After a major repair has been made to an aircraft engine, an FAA Form 337 must be filled out describing the repair.

There must be an entry made in the engine maintenance record that references the Form 337 by its date.

8250. A ground incident that results in propeller sudden stoppage would require a crankshaft runout inspection. What publication would be used to obtain crankshaft runout tolerance?

A—Current manufacturer's maintenance instructions.
B—Type Certificate Data Sheet.
C—AC 43.13-1A, Acceptable Methods, Techniques, and Practices Aircraft Inspection and Repair.

All of the dimensions, limits, and tolerances for a particular engine are included in the engine manufacturer's maintenance instructions for the particular engine.

8251. Select the Airworthiness Directive applicability statement which applies to an IVO-355 engine, serial number T8164, with 2,100 hours' total time and 300 hours since rebuilding.

A—Applies to all IVO-355 engines, serial numbers T8000 through T8300, having less than 2,400 hours' total time.
B—Applies to all IVO-355 engines, serial numbers T8000 through T8900 with 2,400 hours or more total time.
C—Applies to all I.O. and TV10-355 engines, all serial numbers regardless of total time or since overhaul.

Choice (A) lists the type of engine (IVO-355), includes the serial number of the engine (T8164 is between T8000 and T8300) and includes the appropriate time in service (2,100 hours is less than 2,400 hours).

Answers

8245 [C] (C01) 14 CFR 65.95	8246 [B] (C01) 14 CFR 43	8247 [A] (C01) 14 CFR 23	8248 [C] (C01) 14 CFR 43.13
8249 [B] (C01) 14 CFR 43.6	8250 [A] (C01) 14 CFR 43.13a	8251 [A] (C01) AMT-G	

8252. What publication contains the mandatory replacement time for parts of a turbine engine?

A—Engine Manufacturer's service instructions.
B—Airworthiness directive issued by the engine manufacturer.
C—Engine Manufacturer's maintenance manual.

The engine manufacturer's service instructions must contain a section entitled "Instructions for Continued Airworthiness." This section contains mandatory replacement times, inspection intervals, and related procedures required for type certification.

8253. How are discharge nozzles in a fuel injected reciprocating engine identified to indicate the flow range?

A—By an identification letter stamped on one of the hexes of the nozzle body.
B—By an identification metal tag attached to the nozzle body.
C—By color codes on the nozzle body.

The discharge nozzles for a fuel-injected reciprocating engine must be matched with regard to their flow rate.

A code letter is used to identify the flow rate of a particular nozzle. This letter is stamped on one of the hexes of the nozzle body.

8254. What section in the instructions for continued airworthiness is FAA approved?

A—Engine maintenance manual or section.
B—Engine overhaul manual or section.
C—Airworthiness limitations section.

Quoting from 14 CFR Part 33, Appendix A (A33.4): "The Airworthiness Limitations section is FAA approved and specifies maintenance required under §§43.16 and 91.403 of the Federal Aviation Regulations unless an alternative program has been FAA approved."

8255. Which of the following conditions is usually not acceptable to any extent in turbine blades?

A—Cracks.
B—Pits.
C—Dents.

Cracks in a turbine blade are cause for rejection of the blade.

8256. (1) Serviceability limits for turbine blades are much more stringent than are those for turbine nozzle vanes.

(2) A limited number of small nicks and dents can usually be permitted in any area of a turbine blade.

Regarding the above statements,

A—both No. 1 and No. 2 are true.
B—neither No. 1 nor No. 2 is true.
C—only No. 1 is true.

Statement (1) is true. Serviceability limits for turbine blades are much more stringent than those for nozzle vanes. This is particularly true for blades in the first stage because of the high temperature involved. The high centrifugal stresses to which turbine blades are subjected, require that the blades be free of cracks in any area and no nicks or dents are allowed to exist in the root area.

Statement (2) is not true. A limited number of small nicks and dents can be permitted, but only in the areas of the blade away from the root area.

8257. Which unit most accurately indicates fuel consumption of a reciprocating engine?

A—Fuel flowmeter.
B—Fuel pressure gauge.
C—Electronic fuel quantity indicator.

The fuel flowmeter is the most accurate of the choices given here to determine the fuel consumption of a reciprocating engine.

The fuel flowmeter can be cross-checked with the electronic fuel quantity indicator to determine its accuracy.

8258. The fuel flowmeter used with a continuous-fuel injection system installed on an aircraft horizontally opposed reciprocating engine measures the fuel pressure drop across the

A—manifold valve.
B—fuel nozzles.
C—metering valve.

The fuel flowmeter used with a continuous-flow fuel injection system installed on a horizontally opposed engine is a pressure gage that measures the pressure drop across the injector nozzles.

Answers

8252 [A] (C01) 14 CFR 23	8253 [A] (C01) AMT-P	8254 [C] (C01) 14 CFR 33	8255 [A] (C01) AMT-P
8256 [C] (C01) AMT-P	8257 [A] (H01) AMT-P	8258 [B] (H01) AMT-SYS	

8259. The principal fault in the pressure type fuel flowmeter indicating system, installed on a horizontally opposed continuous-flow fuel injected aircraft reciprocating engine, is that a plugged fuel injection nozzle will cause a

A—normal operation indication.
B—lower than normal fuel flow indication.
C—higher than normal fuel flow indication.

The fuel flowmeter used with a continuous-flow fuel injection system installed on a horizontally opposed engine is a pressure gage that measures the pressure drop across the injector nozzles. If one of the nozzles should become plugged, the pressure drop across it would increase and, even though the actual flow has decreased, the flowmeter indicates a higher-than-normal flow.

8260. Motor driven impeller and turbine fuel flow transmitters are designed to transmit data

A—using aircraft electrical system power.
B—mechanically.
C—by fuel pressure.

The mass-flow fuel flowmeter using a motor-driven impeller and turbine, transmit their data using 115-volt AC electrical power from the aircraft electrical system.

8261. The fuel-flow indicator rotor and needle for a motor-impeller and turbine indicating system is driven by

A—an electrical signal.
B—direct coupling to the motor shaft.
C—a mechanical gear train.

The indicator rotor and needle of a mass-flow fuel flowmeter are driven from the transmitter by an electrical signal.

8262. On a twin-engine aircraft with fuel-injected reciprocating engines, one fuel-flow indicator reads considerably higher than the other in all engine operating configurations. What is the probable cause of this indication?

A—Carburetor icing.
B—One or more fuel nozzles are clogged.
C—Alternate air door stuck open.

The fuel flowmeter indicator used in a continuous-flow fuel-injection system is a pressure gauge that reads the pressure drop across the injector nozzles.

A clogged nozzle will cause an increase in flow indication, even though the actual flow has decreased.

8263. The fuel-flow indication system used with many fuel-injected opposed engine airplanes utilizes a measure of

A—fuel flow volume.
B—fuel pressure.
C—fuel flow mass.

The fuel flow indicating system used in the continuous-flow fuel injection system, installed on horizontally opposed engines, is actually a pressure gauge that reads the pressure drop across the injector nozzles. The pressure drop across the nozzles is a function of the volume of fuel flowing through them.

8264. In addition to fuel quantity, a computerized fuel system (CFS) with a totalizer-indicator provides indication of how many of the following?

1. Fuel flow rate.
2. Fuel used since reset or initial start-up.
3. Fuel time remaining at current power setting.
4. Fuel temperature.

A—Two.
B—Three.
C—Four.

The Computerized Fuel System (CFS) provides the pilot with fuel flow information in pounds per hour or gallons per hour; gallons or pounds remaining; time remaining for flight at the current power setting; and gallons used from the initial engine start-up.

8265. The fuel-flow indication data sent from motor driven impeller and turbine, and motorless type fuel flow transmitters is a measure of

A—fuel mass-flow.
B—fuel volume-flow.
C—engine burner pressure drop.

The fuel flow indication data sent from a motor-driven impeller and turbine, and a motorless fuel flow transmitter are measures of mass fuel flow.

These fuel flowmeters take into consideration both the volume and density of the fuel.

8266. In an aircraft equipped with a pressure-drop type fuel-flow indicating system, if one of the injector nozzles becomes restricted, this would cause a decrease in fuel flow with

A—a decreased fuel flow indication on the gauge.
B—an increased fuel flow indication on the gauge.
C—no change in fuel flow indication on the gauge.

Answers

8259 [C] (H01) AMT-P	8260 [A] (H01) AMT-SYS	8261 [A] (H01) AMT-SYS	8262 [B] (H01) AMT-P
8263 [B] (H01) AMT-P	8264 [B] (H01) AMT-P	8265 [A] (H01) AMT-P	8266 [B] (H01) AMT-P

If one of the injector nozzles becomes restricted, the pressure drop across it will increase, and the pressure-drop type flow indicator will show an increase in fuel flow, but there will be a decrease in the actual fuel flow.

8267. A manifold pressure gauge is designed to

A—maintain constant pressure in the intake manifold.
B—indicate differential pressure between the intake manifold and atmospheric pressure.
C—indicate absolute pressure in the intake manifold.

A manifold pressure gauge is an absolute pressure gauge used to measure the absolute pressure (the pressure referenced from zero pressure, or a vacuum) that exists inside the induction system of a reciprocating engine.

8268. The purpose of an exhaust gas analyzer is to indicate the

A—brake specific fuel consumption.
B—fuel/air ratio being burned in the cylinders.
C—temperature of the exhaust gases in the exhaust manifold.

The exhaust-gas analyzer is used to indicate the fuel-air ratio being burned inside the cylinders.

The modern EGT system measures the temperature of the exhaust gases as they leave the cylinders. This gives an indirect, or relative, indication of the mixture being burned.

8269. Which of the following types of electric motors are commonly used in electric tachometers?

A—Direct current, series-wound motors.
B—Synchronous motors.
C—Direct current, shunt-wound motors.

Electric tachometers use a three-phase permanent-magnet generator turned by the engine or helicopter transmission. Tachometer generators produce three-phase alternating current whose frequency is determined by the speed the permanent-magnet rotor is turned.

Inside the instrument case there is a synchronous motor that spins at the same speed as the magnet in the generator.

A magnetic-drag-indicating system converts the rotational speed of the synchronous motor into angular deflection of the tachometer pointer.

8270. Where are the hot and cold junctions located in an engine cylinder temperature indicating system?

A—Both junctions are located at the instrument.
B—Both junctions are located at the cylinder.
C—The hot junction is located at the cylinder and the cold junction is located at the instrument.

A thermocouple, such as is used in a cylinder-head temperature system, is an electrical generator made of two dissimilar metal wires connected together to form a loop. The two points at which the wires touch are called junctions.

The voltage produced by a thermocouple is determined by the temperature difference between the two junctions.

The hot (or measuring) junction is located at the cylinder head. It may be imbedded in a gasket under the spark plug, or it may be in a bayonet held tight against the cylinder head by a spring.

The cold junction is in the case of the instrument.

8271. Basically, the indicator of a tachometer system is responsive to change in

A—current flow.
B—frequency.
C—voltage.

This question refers to the three-phase AC tachometer generator and synchronous motor system.

An AC tachometer generator produces a voltage whose frequency is proportional to the engine speed. The variable frequency alternating current produced by the tachometer generator drives a synchronous motor inside the instrument case.

The pointer of a magnetic drag tachometer is driven by the synchronous motor. It moves across the indicator dial to indicate the speed of the generator rotor.

8272. Which statement is correct concerning a thermocouple-type temperature indicating instrument system?

A—It is a balanced-type, variable resistor circuit.
B—It requires no external power source.
C—It usually contains a balancing circuit in the instrument case to prevent fluctuations of the system voltage from affecting the temperature reading.

A thermocouple-type temperature-indicating system is self-contained and requires no external power.

The voltage generated by a thermocouple is determined by the difference in temperature between the hot (measuring) junction at the cylinder head and the cold (reference) junction inside the instrument case.

Answers

8267 [C] (H02) AMT-P	8268 [B] (H02) AMT-P	8269 [B] (H02) AMT-SYS	8270 [C] (H02) AMT-SYS
8271 [B] (H02) AMT-SYS	8272 [B] (H02) AMT-P		

8273. Which statement is true regarding a thermocouple-type cylinder head temperature measuring system?

A—The resistance required for cylinder head temperature indicators is measured in farads.
B—The voltage output of a thermocouple system is determined by the temperature difference between the two ends of the thermocouple.
C—When the master switch is turned on, a thermocouple indicator will move off-scale to the low side.

The voltage output of a thermocouple is determined by the difference in the temperature of the hot (measuring) junction at the cylinder head and the cold (reference) junction that is inside the instrument case.

8274. What basic meter is used to indicate cylinder head temperature in most aircraft?

A—Electrodynamometer.
B—Galvanometer.
C—Thermocouple-type meter.

We normally associate thermocouples with the measurement of cylinder-head temperature. But this question asks for the basic meter used to indicate cylinder-head temperature.

A galvanometer is the same as a D'Arsonval current-measuring instrument. This is the basic meter used to indicate cylinder-head temperature.

The thermocouple is the device that generates a voltage proportional to the temperature of the cylinder head.

8275. Which of the following is a primary engine instrument?

A—Tachometer.
B—Fuel flowmeter.
C—Airspeed indicator.

If we consider the instruments required by the FAA for certificated aircraft as "primary" engine instruments, the tachometer is the only one of these instruments that is a primary engine instrument.

8276. A complete break in the line between the manifold pressure gauge and the induction system will be indicated by the gauge registering

A—prevailing atmospheric pressure.
B—zero.
C—lower than normal for conditions prevailing.

If the break is complete, the engine will have no effect on the instrument, and the manifold pressure gauge will indicate the prevailing atmospheric pressure at all engine speeds.

8277. Engine oil temperature gauges indicate the temperature of the oil

A—entering the oil cooler.
B—entering the engine.
C—in the oil storage tank.

The temperature indicated on the oil-temperature gauge is the temperature of the oil as it enters the engine.

Oil-inlet temperature gives an indication of the efficiency of the oil-cooling system. It lets the pilot know whether or not the oil entering the engine is too hot for adequate heat removal.

8278. Why do helicopters require a minimum of two synchronous tachometer systems?

A—One indicates engine RPM and the other tail rotor RPM.
B—One indicates main rotor RPM and the other tail rotor RPM.
C—One indicates engine RPM and the other main rotor RPM.

Helicopters use a dual tachometer. One tachometer shows the speed of the engine and the other shows the speed of the main rotor.

The ratio between the tachometer generator drive for the engine and that for the main rotor is such that when the clutch for the main rotor is fully engaged, the needles for the two tachometers are "married"; one is directly on top of the other.

When the needles are "split," the main rotor clutch is not fully engaged.

8279. If the thermocouple leads were inadvertently crossed at installation, what would the cylinder temperature gauge pointer indicate?

A—Normal temperature for prevailing condition.
B—Moves off-scale on the zero side of the meter.
C—Moves off-scale on the high side of the meter.

The cylinder-head temperature gauge is a current-measuring instrument that measures the current produced in the thermocouple.

If the thermocouple leads were inadvertently crossed at installation, the indicator would read backward, which would cause the pointer to peg out on the low side of the scale.

Answers

8273 [B] (H02) AMT-P	8274 [B] (H02) AMT-SYS	8275 [A] (H02) AMT-P	8276 [A] (H02) AMT-P
8277 [B] (H02) AMT-P	8278 [C] (H02) AMT-SYS	8279 [B] (H02) AMT-SYS	

8280. A common type of electrically operated oil temperature gauge utilizes

A—either a wheatstone bridge or ratiometer circuit.
B—a thermocouple type circuit.
C—vapor pressure and pressure switches.

The most generally used electrical oil temperature gauges for aircraft engines uses either a Wheatstone bridge or a ratiometer circuit.

A temperature probe consisting of a coil of thin wire encased in a stainless steel housing, senses the temperature of the oil. The resistance of the wire changes as the temperature changes. This change of resistance is indicated on the instrument as a change in temperature.

8281. The indication on a thermocouple-type cylinder head temperature indicator is produced by

A—resistance changes in two dissimilar metals.
B—a difference in the voltage between two dissimilar metals.
C—a current generated by the temperature difference between dissimilar metal hot and cold junctions.

The indication on a thermocouple-type cylinder-head temperature indicator is produced by a temperature difference between the two junctions (the measuring junction and the reference junction).

The temperature difference between these two junctions causes a current to flow through the thermocouple system that is proportional to the temperature difference between the junctions.

8282. (1) Powerplant instrument range markings show whether the current state of powerplant operation is normal, acceptable for a limited time, or unauthorized.

(2) Powerplant instrument range markings are based on installed engine operating limits which may not exceed (but are not necessarily equal to) those limits shown on the engine Type Certificate Data Sheet.

Regarding the above statements,

A—both No. 1 and No. 2 are true.
B—neither No. 1 nor No. 2 is true.
C—only No. 1 is true.

Statement (1) is true. Powerplant instrument range markings show whether the current state of powerplant operation is normal (green arc), acceptable for a limited time (yellow arc), or unauthorized (red line).

Statement (2) is also true. Powerplant range markings are based on the installed engine operating limits which are found in the Aircraft Type Certificate Data Sheets. These limits may not exceed, but are not necessarily the same as those shown in the Engine Type Certificate Data Sheets.

8283. Thermocouple leads

A—may be installed with either lead to either post of the indicator.
B—are designed for a specific installation and may not be altered.
C—may be repaired using solderless connectors.

Thermocouple leads must have a specific resistance for a given installation, and therefore they should not be altered.

8284. (1) Engine pressure ratio (EPR) is a ratio of the exhaust gas pressure to the engine inlet air pressure, and indicates the thrust produced.

(2) Engine pressure ratio (EPR) is a ratio of the exhaust gas pressure to the engine inlet air pressure, and indicates volumetric efficiency.

Regarding the above statements,

A—only No. 1 is true.
B—only No. 2 is true.
C—both No. 1 and No. 2 are true.

Statement (1) is true. Engine pressure ratio (EPR) is a ratio of the turbine discharge pressure to the engine inlet pressure. It relates to the amount of thrust the engine is producing.

Statement (2) is not true. The EPR has nothing to do with indicating the volumetric efficiency of an engine.

8285. What unit in a tachometer system sends information to the indicator?

A—The three-phase ac generator.
B—The two-phase ac generator.
C—The synchronous motor.

The tachometer system mentioned in this question uses a three-phase AC generator mounted on the engine to drive a synchronous motor inside the tachometer indicator.

Answers

8280 [A] (H02) AMT-P	8281 [C] (H02) AMT-P	8282 [A] (H02) AMT-P	8283 [B] (H02) AMT-P
8284 [A] (H02) AMT-P	8285 [A] (H02) AMT-SYS		

8286. (1) Generally, when a turbine engine indicates high EGT for a particular EPR (when there is no significant damage), it means that the engine is out of trim.

(2) Some turbine-powered aircraft use RPM as the primary indicator of thrust produced, others use EPR as the primary indicator.

Regarding the above statements,

A—only No. 1 is true.
B—only No. 2 is true.
C—both No. 1 and No. 2 are true.

Statement (1) is true. If the EGT is high for a particular EPR when there is no significant damage, the engine is out of trim and the fuel control should be adjusted.

Statement (2) is also true. Turbine engines with centrifugal compressors use RPM as a primary indicator of thrust while axial-flow engines use EPR as a primary thrust indicator.

8287. Engine pressure ratio is determined by

A—multiplying engine inlet total pressure by turbine outlet total pressure.
B—dividing turbine outlet total pressure by engine inlet total pressure.
C—dividing engine inlet total pressure by turbine outlet total pressure.

Engine pressure ratio (EPR) is the ratio of the turbine-discharge pressure to the compressor-inlet pressure. It is used as a measure of the thrust produced by an axial-flow turbine engine.

A ratio is found by dividing one term by the other. In this instance, the turbine-outlet total pressure is divided by the engine-inlet (compressor-inlet) total pressure.

8288. Jet engine thermocouples are usually constructed of

A—chromel-alumel.
B—iron-constantan.
C—alumel-constantan.

The thermocouples used to measure EGT or TIT of a turbojet engine are made of chromel and alumel wire.

These temperatures are much higher than the cylinder-head temperature of a reciprocating engine that uses thermocouples made of iron and constantan or copper and constantan.

8289. Which of the following instrument discrepancies require replacement of the instrument?

1. Red line missing from glass.
2. Glass cracked.
3. Case paint chipped.
4. Will not zero out.
5. Pointer loose on shaft.
6. Mounting screw loose.
7. Leaking at line B nut.
8. Fogged.

A—2, 3, 7, 8.
B—2, 4, 5, 8.
C—1, 2, 4, 7.

1. *A red line missing from the instrument glass will not require replacement of the instrument.*
2. *A cracked glass will require instrument replacement.*
3. *Case paint chipped will not require instrument replacement.*
4. *If the instrument will not zero out, it will have to be replaced.*
5. *A pointer loose on the shaft will require instrument replacement.*
6. *Loose mounting screws will not require instrument replacement.*
7. *A leaking B-nut will not require instrument replacement.*
8. *If the instrument is fogged, there is probably a vent problem and the instrument will have to be replaced.*

8290. A Bourdon-tube instrument may be used to indicate

1. pressure.
2. temperature.
3. position.
4. quantity.

A—1 and 2.
B—1 and 3.
C—2 and 4.

A Bourdon tube measures only pressure. But in a Bourdon tube temperature-measuring instrument, the Bourdon tube is connected by a sealed capillary tube to a bulb containing methyl chloride.

The bulb is placed in the area where the temperature is to be measured. As the temperature changes, the pressure of the methyl chloride changes.

Answers
8286 [C] (H02) AMT-P
8287 [B] (H02) AMT-P
8288 [A] (H02) AMT-SYS
8289 [B] (H02) AMT-SYS
8290 [A] (H02) AMT-P

8291. An indication of unregulated power changes that result in continual drift of manifold pressure indication on a turbosupercharged aircraft engine is known as?

A—Overshoot.
B—Waste gate fluctuation.
C—Bootstrapping.

Bootstrapping is a transient increase in engine power that causes the turbocharger to speed up, which in turn causes the engine to produce more power. Bootstrapping is indicated by a continual drift in the manifold pressure indication.

8292. Which of the following instrument conditions is acceptable and does NOT require immediate correction?

1. Red line missing.
2. Pointer loose on shaft.
3. Glass cracked.
4. Mounting screws loose.
5. Case paint chipped.
6. Leaking at line B nut.
7. Will not zero out.
8. Fogged.

A—1.
B—4.
C—5.

The only one of these conditions that is acceptable and would not require immediate correction is the chipped case paint.

8293. A change in engine manifold pressure has a direct effect on the

A—piston displacement.
B—compression ratio.
C—mean effective cylinder pressure.

The mean effective pressure inside the cylinder of a reciprocating engine is the average pressure in the cylinder during the power stroke.

This average pressure is affected by both the manifold pressure (the pressure of the air as it enters the cylinder) and the compression ratio of the engine.

8294. What instrument on a gas turbine engine should be monitored to minimize the possibility of a "hot" start?

A—RPM indicator.
B—Turbine inlet temperature.
C—Torquemeter.

Turbine-inlet temperature (TIT) or exhaust-gas temperature (EGT) of a gas turbine engine is monitored to determine whether the start is normal or if it is a hot start.

8295. In regard to using a turbine engine oil analysis program, which of the following is NOT true?

A—Generally, an accurate trend forecast may be made after an engine's first oil sample analysis.
B—It is best to start an oil analysis program on an engine when it is new.
C—A successful oil analysis program should be run over an engine's total operating life so that normal trends can be established.

An oil analysis program is a trend indicating system. A sample of the oil is taken when the engine is new and the parts-per-million of each of a number of elements is recorded. As subsequent samples are tested on a regular basis, the trend of the growth of the trace elements are recorded. When the growth of any element is far more rapid than it should be, the engine operator is warned of possible impending trouble.

Analysis of the first sample is only a starting point and cannot give an accurate forecast alone.

8296. On an aircraft turbine engine, operating at a constant power, the application of engine anti-icing will result in

A—noticeable shift in EPR.
B—a false EPR reading.
C—an increase in EPR.

When an engine is operating at a constant power and anti-ice is selected, a slight rise in EGT indicates the system is operating properly. The EPR and RPM will shift noticeably because of the change in the mass of air delivered to the combustor.

8297. Engine pressure ratio is the total pressure ratio between the

A—aft end of the compressor and the aft end of the turbine.
B—front of the compressor and the rear of the turbine.
C—front of the engine inlet and the aft end of the compressor.

Engine pressure ratio (EPR) is the ratio of the turbine-discharge total pressure (measured at the aft end of the turbine) to the compressor-inlet total pressure (measured at the front end of the compressor).

Total pressure is the pressure a column of moving air has when it is stopped.

Answers

8291 [C] (H02) 14 CFR 65.81	8292 [C] (H02) AMT-SYS	8293 [C] (H02) AMT-P	8294 [B] (H02) AMT-P
8295 [A] (H02) AMT-P	8296 [A] (H02) AMT-P	8297 [B] (H02) AMT-P	

8298. What would be the possible cause if a gas turbine engine has high exhaust gas temperature, high fuel flow, and low RPM at all engine power settings?

A—Fuel control out of adjustment.
B—Loose or corroded thermocouple probes for the EGT indicator.
C—Turbine damage or loss of turbine efficiency.

A damaged or dirty turbine can cause the engine to have a high exhaust-gas temperature, a high fuel flow and a low RPM at all power settings.

8299. What is the primary purpose of the tachometer on an axial-compressor turbine engine?

A—Monitor engine RPM during cruise conditions.
B—It is the most accurate instrument for establishing thrust settings under all conditions.
C—Monitor engine RPM during starting and to indicate overspeed conditions.

The RPM of an axial-flow gas turbine engine is not used as an indicator of thrust, but RPM is important during the engine starting procedure and during high-power operation to prevent overspeeding the engine.

8300. The engine pressure ratio (EPR) indicator is a direct indication of

A—engine thrust being produced.
B—pressure ratio between the front and aft end of the compressor.
C—ratio of engine RPM to compressor pressure.

In an axial-flow gas turbine engine, the engine pressure ratio (EPR) varies directly with the thrust the engine is producing.

8301. The exhaust gas temperature (EGT) indicator on a gas turbine engine provides a relative indication of the

A—exhaust temperature.
B—temperature of the exhaust gases as they pass the exhaust cone.
C—turbine inlet temperature.

The temperature of the gases entering the turbine is very difficult to measure.

Because of this, exhaust-gas temperature which is easy to measure, is used to give an indication of the turbine-inlet temperature.

8302. What instrument indicates the thrust of a gas turbine engine?

A—Exhaust gas temperature indicator.
B—Turbine inlet temperature indicator.
C—Engine pressure ratio indicator.

In an axial-flow gas turbine engine, the engine pressure ratio (EPR) varies directly with the thrust the engine is producing.

8303. In a turbine engine, where is the turbine discharge pressure indicator sensor located?

A—At the aft end of the compressor section.
B—At a location in the exhaust cone that is determined to be subjected to the highest pressures.
C—Immediately aft of the last turbine stage.

A turbine-discharge pressure-indicator sensor is located immediately aft of the last turbine stage.

The turbine-discharge pressure is used with the compressor-inlet pressure to find the engine pressure ratio (the EPR).

8304. In what units are turbine engine tachometers calibrated?

A—Percent of engine RPM.
B—Actual engine RPM.
C—Percent of engine pressure ratio.

Gas turbine engine tachometers are calibrated in percent RPM of the compressor.

It is much more important that the pilot know whether or not he is operating the engine at its peak speed or an overspeed condition, than it is for him to know specific RPMs.

8305. Instruments that provide readings of low or negative pressure, such as manifold pressure gauges, are usually what type?

A—Vane with calibrated spring.
B—Bourdon tube.
C—Diaphragm or bellows.

Because of their sensitivity, instruments which measure low or negative pressures are normally of the diaphragm or bellows type.

Answers

8298 [C] (H02) AMT-P	8299 [C] (H02) AMT-P	8300 [A] (H02) AMT-P	8301 [C] (H02) AMT-P
8302 [C] (H02) AMT-P	8303 [C] (H02) AMT-P	8304 [A] (H02) AMT-P	8305 [C] (H02) AMT-SYS

8306. Instruments that measure relatively high fluid pressures, such as oil pressure gauges, are usually what type?

A—Vane with calibrated spring.
B—Bourdon tube.
C—Diaphragm or bellows.

Bourdon tube instruments are normally used to measure such pressures as engine oil pressure and hydraulic system pressure.

8307. The RPM indication of a synchronous ac motor-tachometer is governed by the generator

A—voltage.
B—current.
C—frequency.

The RPM indication of a synchronous AC tachometer is governed by the generator frequency.

The tachometer generator used in this type of system is a permanent magnet generator whose output frequency is determined by the speed it is turned.

8308. The EGT gauge used with reciprocating engines is primarily used to furnish temperature readings in order to

A—obtain the best mixture setting for fuel efficiency.
B—obtain the best mixture setting for engine cooling.
C—prevent engine overtemperature.

The exhaust gas temperature indicating system on a reciprocating engine is used primarily to obtain the best fuel-air mixture setting for fuel efficiency.

After the airplane is trimmed for cruise flight, the mixture is leaned until the EGT peaks. Then it is enriched until the EGT drops a specified amount. This indication lets the pilot knows he is operating with a mixture that is slightly on the rich side of the most efficient mixture for this power setting.

8309. A red triangle, dot, or diamond mark on an engine instrument face or glass indicates

A—the maximum operating limit for all normal operations.
B—the maximum limit for high transients such as starting.
C—a restricted operating range.

A red triangle, dot, or diamond mark on an engine instrument face or glass, indicates the maximum limit for high transients such as starting.

8310. Which of the following fire detectors are commonly used in the power section of an engine nacelle?

A—CO detectors.
B—Smoke detectors.
C—Rate-of-temperature-rise detectors.

Watch this type of question. The question asks for the type of fire detector that would be used.

Of the choices given here, only one is a fire detector. The rate-of-temperature-rise detector is a fire detector. The CO (carbon monoxide) detector, and smoke detector are not fire detectors.

8311. What is the function of a fire detection system?

A—To discharge the powerplant fire-extinguishing system at the origin of the fire.
B—To activate a warning device in the event of a powerplant fire.
C—To identify the location of a powerplant fire.

A fire detection system activates a warning device in the event of a powerplant fire, but it does not discharge the fire-extinguishing agent.

A switch must be actuated by the pilot or flight engineer to discharge the fire-extinguishing agent.

8312. (Refer to Figure 2.) Determine the fire-extinguisher container pressure limits when the temperature is 75°F.

A—326 minimum and 415 maximum.
B—330 minimum and 419 maximum.
C—338 minimum and 424 maximum.

CONTAINER PRESSURE VERSUS TEMPERATURE

TEMPERATURE °F	CONTAINER PRESSURE (PSIG)	
	MINIMUM	MAXIMUM
-40	60	145
-30	83	165
-20	105	188
-10	125	210
0	145	230
10	167	252
20	188	275
30	209	295
40	230	317
50	255	342
60	284	370
70	319	405
80	356	443
90	395	483
100	438	523

Figure 2

To work this problem, we must interpolate. Since 75°F is halfway between 70° and 80°F, we must find a minimum and maximum pressure that is halfway between the pressures for 70° and 80°F. The minimum pressure is 338 psig, and the maximum pressure is 424 psig.

Answers

8306 [B] (H02) AMT-SYS	8307 [C] (H02) AMT-P	8308 [A] (H02) AMT-P	8309 [B] (H02) AC 20-88A
8310 [C] (I01) AMT-P	8311 [B] (I01) AMT-P	8312 [C] (I01) AMT-P	

8313. How are most aircraft turbine engine fire-extinguishing systems activated?

A—Electrically discharged cartridges.
B—Manual remote control valve.
C—Pushrod assembly.

Most aircraft-turbine-engine fire-extinguishing systems use a high-rate discharge (HRD) bottle that contains some type of Freon pressurized with nitrogen.

Electrically operated squibs (explosive charges) rupture the metal seal on the HRD bottle and discharge the fire-extinguishing agent.

8314. How does carbon dioxide (CO_2) extinguish an aircraft engine fire?

A—Contact with the air converts the liquid into snow and gas which smothers the flame.
B—By lowering the temperature to a point where combustion will not take place.
C—The high pressure spray lowers the temperature and blows out the fire.

When a carbon dioxide fire extinguisher is discharged into an aircraft engine, liquid CO_2 converts into a gas and snow which displaces the oxygen in the immediate area of the fire and smothers the flame. The fire goes out when it is deprived of oxygen.

8315. What retains the nitrogen charge and fire-extinguishing agent in a high rate of discharge (HRD) container?

A—Breakable disk and fusible disk.
B—Pressure switch and check tee valve.
C—Pressure gauge and cartridge.

The high-rate discharge containers are sealed with a breakable disk that is cut by an explosive cartridge when the agent discharge switch is energized.

There is also a fusible disk that will rupture and release the agent if the temperature surrounding the container becomes excessive.

If the bottle is discharged by the normal method, a yellow indicator disk will blow out, but if it is discharged by an overtemperature condition, the red indicator disk will blow out.

8316. A continuous-loop fire detector is what type of detector?

A—Spot detector.
B—Overheat detector.
C—Rate-of-temperature-rise detector.

A continuous-loop fire detection system is a form of overheat detector made in the form of a loop installed around the engine compartment.

It initiates a fire warning when a fire or overheat condition changes the electrical characteristics of the heat-sensitive material in the loop.

8317. What is the operating principle of the spot detector sensor in a fire detection system?

A—Resistant core material that prevents current flow at normal temperatures.
B—A conventional thermocouple that produces a current flow.
C—A bimetallic thermoswitch that closes when heated to a high temperature.

The spot detector sensor in a fire detection system is a bimetallic thermal switch between two loops of wire.

When a high temperature occurs at the detector, the switch closes and completes the circuit between the two loops. This initiates the fire warning signal.

8318. How is the fire-extinguishing agent distributed in the engine section?

A—Spray nozzles and fluid pumps.
B—Nitrogen pressure and slinger rings.
C—Spray nozzles and perforated tubing.

The fire-extinguishing agent in a reciprocating engine nacelle is usually distributed through perforated tubing and in turbine engines through spray-type discharge nozzles.

8319. Which of the following is the safest fire-extinguishing agent to use from a standpoint of toxicity and corrosion hazards?

A—Dibromodifluoromethane (Halon 1202).
B—Bromochlorodifluoromethane (Halon 1211).
C—Bromotrifluoromethane (Halon 1301).

Halon 1301 (bromotrifluoromethane) is the safest fire extinguishing agent listed from the standpoints of both toxicity and corrosion hazards.

8320. Which of the following is NOT used to detect fires in reciprocating engine nacelles?

A—Smoke detectors.
B—Rate-of-temperature-rise detectors.
C—Flame detectors.

Smoke detectors are not normally used to detect fires in a reciprocating engine nacelle. Smoke is often produced when the engine starts due to oil in the cylinders. This is not necessarily an indication of fire.

Answers

8313 [A] (I01) AMT-P	8314 [A] (I01) AMT-P	8315 [A] (I01) AMT-P	8316 [B] (I01) AMT-SYS
8317 [C] (I01) AMT-P	8318 [C] (I01) AMT-P	8319 [C] (I01) AMT-SYS	8320 [A] (I01) AMT-P

8321. What is the principle of operation of the continuous-loop fire detector system sensor?

A—Fuse material which melts at high temperatures.
B—Core resistance material which prevents current flow at normal temperatures.
C—A bimetallic thermoswitch which closes when heated to a high temperature.

The core material that serves as an insulator in a continuous-loop fire detection system prevents current flowing between the conductors under normal operating conditions.

But, when a fire occurs, the loop is heated and the resistance of the insulating material becomes low enough to allow current to pass between the two conductors and initiate a fire warning.

8322. The most satisfactory extinguishing agent for a carburetor or intake fire is

A—carbon dioxide.
B—dry chemical.
C—methyl bromide.

Carbon dioxide, when properly used on an induction system fire, will put out the fire and not damage the engine as methyl bromide or dry chemicals would.

8323. The explosive cartridge in the discharge valve of a fire-extinguisher container is

A—a life-dated unit.
B—not a life-dated unit.
C—mechanically fired.

The service life of a fire-extinguisher discharge cartridge is recommended by the manufacturer, usually in terms of hours.

Some cartridges have a service life of 5,000 hours.

8324. Why does one type of Fenwal fire detection system use spot detectors wired in parallel between two separate circuits?

A—To provide an installation that is equal to two separate systems: a primary system and a secondary, or back-up system.
B—So that a double fault may exist in the system without sounding a false alarm.
C—So that a single fault may exist in the system without sounding a false alarm.

A fire warning is initiated when the two-terminal Fenwal spot detectors complete the circuit between the two loops of wire.

The system can withstand one fault, either an electrical open circuit or a short to ground without sounding a false fire warning.

8325. Which of the following fire detection systems measures temperature rise compared to a reference temperature?

A—Thermocouple.
B—Thermal switch.
C—Lindberg continuous element.

The thermocouple system uses active thermocouples in the fire area and a reference thermocouple (reference junction) enclosed in a dead air space between two blocks of insulation material.

The thermocouple system senses an excessive rate of temperature rise to indicate the presence of a fire.

8326. The pulling out (or down) of an illuminated fire handle in a typical large jet aircraft fire protection system commonly accomplishes what events?

A—Closes all firewall shutoff valves, disconnects the generator, and discharges a fire bottle.
B—Closes fuel shutoff, closes hydraulic shutoff, disconnects the generator field, and arms the fire-extinguishing system.
C—Closes fuel shutoff, closes hydraulic shutoff, closes the oxygen shutoff, disconnects the generator field, and arms the fire-extinguishing system.

When the illuminated fire handle on a jet aircraft is pulled out or down, it closes the fuel shutoff, closes the hydraulic fluid shutoff, disconnects the generator field, and arms the fire-extinguishing system.

It does not discharge the fire-extinguishing agent.

8327. A fire detection system operates on the principle of a buildup of gas pressure within a tube proportional to temperature. Which of the following systems does this statement define?

A—Kidde continuous-loop system.
B—Lindberg continuous-element system.
C—Thermal switch system.

The Lindberg continuous-element fire detection system operates on a buildup of gas pressure from gas released by the element inside the tube when it is heated.

Increased gas pressure moves a diaphragm and closes a switch to signal a fire or overheat condition.

Answers

8321 [B] (I01) AMT-P
8322 [A] (I01) AMT-P
8323 [A] (I01) AMT-SYS
8324 [C] (I01) AMT-SYS
8325 [A] (I01) AMT-P
8326 [B] (I01) AMT-P
8327 [B] (I01) AMT-P

8328. The fire detection system that uses a single wire surrounded by a continuous string of ceramic beads in a tube is the

A—Fenwal system.
B—Kidde system.
C—thermocouple system.

The Fenwal continuous-loop fire detection system uses a string of ceramic beads in a tube to hold the center conductor insulated from the outer tube.

The conductivity of these beads increases as the beads get hot.

8329. The fire detection system that uses two wires imbedded in a ceramic core within a tube is the

A—Fenwal system.
B—Lindberg system.
C—Kidde system.

The Kidde continuous-loop fire detection system uses two wires embedded in a ceramic insulator whose conductivity increases as it gets hot.

The insulator and the two wires are housed in a continuous metal tube.

8330. A fuel or oil fire is defined as a

A—class B fire.
B—class A fire.
C—class C fire.

A class-A fire involves ordinary combustible materials such as wood, cloth, paper, upholstery materials, etc.

A class-B fire involves flammable petroleum products or other flammable or combustible liquids.

A class-C fire involves energized electrical equipment.

A class-D fire is a fire in a flammable metal.

8331. A fire detection system that operates on the rate-of-temperature rise is a

A—continuous-loop system.
B—thermocouple system.
C—thermal switch system.

A thermocouple fire detection system operates on the rate-of-temperature-rise.

Active thermocouples are located in the fire zone. A reference thermocouple is enclosed in a dead-air space between two blocks of insulating material. As long as the active and reference thermocouples are the same temperature, the system is inactive but in the event of a fire, the temperature of the active thermocouple rises much faster than that of the reference thermocouple so, the system signals a fire.

8332. A fire involving energized electrical equipment is defined as a

A—class B fire.
B—class D fire.
C—class C fire.

A class-A fire involves ordinary combustible materials such as wood, cloth, paper, upholstery materials, etc.

A class-B fire involves flammable petroleum products or other flammable or combustible liquids.

A class-C fire involves energized electrical equipment.

A class-D fire is a fire in a flammable metal.

8333. Two continuous-loop fire detection systems that will not test due to a broken detector element are the

A—Kidde system and the Lindberg system.
B—Kidde system and the Fenwal system.
C—thermocouple system and the Lindberg system.

The Kidde and the Fenwal fire detection systems are the two continuous-loop fire detection systems that will not test good if the detector element is broken.

Both systems, however, will detect a fire even though they have a broken wire and test as being not good.

8334. In a fixed fire-extinguishing system, there are two small lines running from the system and exiting overboard. These line exit ports are covered with a blowout type indicator disc. Which of the following statements is true?

A—When the red indicator disc is missing, it indicates the fire-extinguishing system has been normally discharged.
B—When the yellow indicator disc is missing, it indicates the fire-extinguishing system has been normally discharged.
C—When the green indicator disc is missing, it indicates the fire-extinguishing system has had a thermal discharge.

The yellow blowout indicator being blown out indicates that the fire extinguisher has been discharged normally.

If the red indicator had been blown out, it would indicate that the extinguisher has been discharged by an overheat (thermal) condition.

8335. The most satisfactory extinguishing agent for an electrical fire is

A—carbon tetrachloride.
B—carbon dioxide.
C—methyl bromide.

Answers

8328 [A] (I01) AMT-SYS	8329 [C] (I01) AMT-SYS	8330 [A] (I01) AMT-P	8331 [B] (I01) AMT-P
8332 [C] (I01) AMT-G	8333 [B] (I01) AMT-P	8334 [B] (I01) AMT-SYS	8335 [B] (I01) AMT-SYS

Carbon dioxide is the best of the fire-extinguishing agents listed here for extinguishing a fire when there are energized electrical wires in the fire.

8336. Which of the following fire detection systems will detect a fire when an element is inoperative but will not test when the test circuit is energized?

A—The Kidde system and the thermocouple system.
B—The Kidde system and the Fenwal system.
C—The thermocouple system and the Lindberg system.

The Kidde and the Fenwal fire detection systems are the two continuous-loop fire detection systems.

Neither of these systems will test (show that they are good) if the detector wire is broken.

Both systems, however, will detect a fire even though they have a broken wire.

8337. Which of the following fire detection systems uses heat in the normal testing of the system?

A—The thermocouple system and the Lindberg system.
B—The Kidde system and the Fenwal system.
C—The thermocouple system and the Fenwal system.

The thermocouple system and the Lindberg fire detection system both use heat to produce a test signal.

The other systems test by measuring continuity of the heat-sensitive element.

8338. After a fire is extinguished, or overheat condition removed in aircraft equipped with a Systron-Donner fire detector, the detection system

A—must be manually reset.
B—automatically resets.
C—sensing component must be replaced.

When a fire that has been detected by a Systron-Donner pneumatic fire detection unit is extinguished and the temperature drops, the hydrogen gas that had been released inside the detection unit is re-absorbed and the pressure in the tube is reduced. This opens the pneumatic switch and the system returns to normal, ready to signal another fire or overheat condition.

8339. The use of water on class D fires

A—is most effective if sprayed in a fine mist.
B—will cause the fire to burn more violently and can cause explosions.
C—has no effect.

A class-D fire is a fire involving a flammable metal. This type of fire requires special handling.

Water used on a magnesium fire will accelerate the burning and can cause an explosion. Special dry powders are available for use whenever metal fires are possible.

8340. For fire detection and extinguishing purposes, aircraft powerplant areas are divided into fire zones based on

A—hot and cold sections of the engine.
B—the volume and smoothness of the airflow through engine compartments.
C—engine type and size.

For fire detection and extinguishing purposes, aircraft powerplant areas are divided into fire zones based on the volume and the smoothness of the airflow.

8341. (Refer to Figure 3.) What are the fire-extinguisher container pressure limits when the temperature is 50°F?

A—425 – 575 PSIG.
B—435 – 605 PSIG.
C—475 – 625 PSIG.

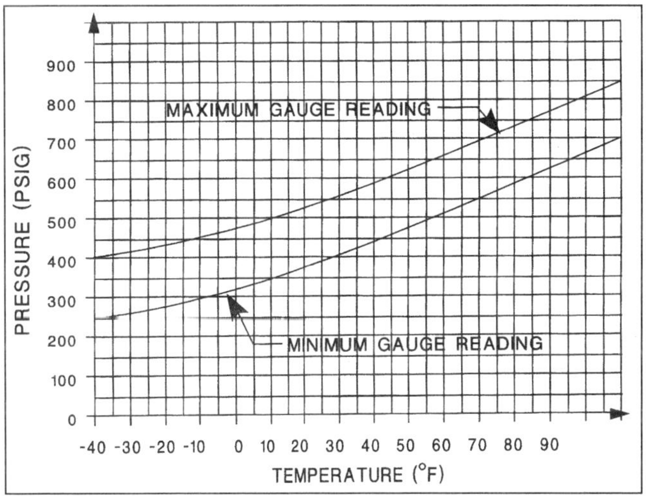

Figure 3

To find the allowable range of fire extinguisher container pressure at 50°F, follow the vertical line for 50°F upward until it intersects the curve for the minimum gauge reading. From this point, follow a horizontal line to the left until it intersects the pressure index. This intersection is at 475 psig.

Continue to follow the 50°F vertical line upward until it intersects the curve for the maximum gauge reading. From this point, follow a horizontal line to the left until it intersects the pressure index. This intersection is at 625 psig.

Answers

8336 [B] (I01) AMT-P
8337 [A] (I01) AMT-P
8338 [B] (I01) AMT-SYS
8339 [B] (I01) AMT-G
8340 [B] (I01) AMT-P
8341 [C] (I01) AMT-P

8342. What device is used to convert alternating current, which has been induced into the loops of the rotating armature of a dc generator, to direct current?

A—A rectifier.
B—A commutator.
C—An inverter.

A commutator is a mechanical switch, or rectifier, that converts the alternating current generated within the rotating armature of a DC generator into direct current as it leaves the generator.

8343. A certain direct current series motor mounted within an aircraft draws more amperes during start than when it is running under its rated load. The most logical conclusion that may be drawn is

A—the starting winding is shorted.
B—the brushes are floating at operating RPM because of weak brush springs.
C—the condition is normal for this type of motor.

A series-wound DC motor draws a very high current as it starts, but when it runs at its normal rate of speed, the current furnished by the source decreases because of the back EMF generated in the motor as it turns.

8344. The stationary field strength in a direct current generator is varied

A—by the reverse-current relay.
B—because of generator speed.
C—according to the load requirements.

The strength of the stationary field of a DC generator is varied according to the load requirement.

As the load increases, the generator voltage tends to drop, but the voltage regulator automatically increases the field current to bring the voltage back high enough to provide the required current to the load.

8345. What type of electric motor is generally used with a direct-cranking engine starter?

A—Direct current, shunt-wound motor.
B—Direct current, series-wound motor.
C—Synchronous motor.

A series-wound DC motor has the highest starting torque of any of the motors listed here. This makes it useful for direct-cranking electric starters.

8346. Upon what does the output frequency of an ac generator (alternator) depend?

A—The speed of rotation and the strength of the field.
B—The speed of rotation, the strength of the field, and the number of field poles.
C—The speed of rotation and the number of field poles.

The frequency of alternating current produced by an AC generator is determined by the number of field poles and the speed of rotation of the rotor.

The frequency, in hertz, of the alternating current produced by a generator is found by multiplying the number of pairs of poles by the RPM divided by 60.

8347. A high surge of current is required when a dc electric motor is first started. As the speed of the motor increases,

A—the counter emf decreases proportionally.
B—the applied emf increases proportionally.
C—the counter emf builds up and opposes the applied emf, thus reducing the current flow through the armature.

When the switch is first closed on a DC motor, there is a high surge of current, but as soon as the motor starts turning, it acts as a generator and builds up a counter EMF that opposes the applied EMF.

The counter EMF decreases the current flow as the armature speeds up.

8348. Alternators (ac generators) that are driven by a constant-speed drive (CSD) mechanism are used to regulate the alternator to a constant

A—voltage output.
B—amperage output.
C—hertz output.

AC generators (alternators) are driven through a constant-speed-drive (CSD) mechanism so they will produce a constant frequency (constant number of cycles per second, or hertz) as the engine speed varies through its normal operating range.

8349. What is used to polish commutators or slip rings?

A—Very fine sandpaper.
B—Crocus cloth or fine oilstone.
C—Aluminum oxide or garnet paper.

Very fine sandpaper is the only abrasive listed here that should be properly used to polish commutators or slip rings.

Crocus cloth and aluminum oxide are conductive and can cause a short circuit between the commutator segments.

Answers

8342 [B] (J01) AMT-G	8343 [C] (J01) AMT-G	8344 [C] (J01) AMT-G	8345 [B] (J01) AMT-P
8346 [C] (J01) AMT-G	8347 [C] (J01) AMT-G	8348 [C] (J01) AMT-G	8349 [A] (J01) AMT-G

8350. If a generator is malfunctioning, its voltage can be reduced to residual by actuating the

A—rheostat.
B—generator master switch.
C—master solenoid.

If a generator malfunctions, its output voltage can be immediately reduced to its residual voltage by opening the generator master switch. This opens the generator field circuit and reduces the generator output to its residual voltage.

Most modern aircraft use a split-type master switch which allows the generator or alternator to be turned off without disconnecting the battery from the aircraft electrical system.

8351. If the points in a vibrator-type voltage regulator stick in the closed position while the generator is operating, what will be the probable result?

A—Generator output voltage will decrease.
B—Generator output voltage will not be affected.
C—Generator output voltage will increase.

If the voltage regulator points stick in the closed position, the electromagnetic pull of the voltage coil cannot open them so the generator output voltage will increase.

8352. Why is a constant-speed drive used to control the speed of some aircraft engine-driven generators?

A—So that the voltage output of the generator will remain within limits.
B—To eliminate uncontrolled surges of current to the electrical system.
C—So that the frequency of the alternating current output will remain constant.

Constant-speed-drive (CSD) units are used between aircraft engines and AC alternators to maintain the frequency of the output voltage constant as the engine RPM changes within the normal operating range.

8353. According to the electron theory of the flow of electricity, when a properly functioning dc alternator and voltage regulating system is charging an aircraft's battery, the direction of current flow through the battery

A—is into the negative terminal and out the positive terminal.
B—is into the positive terminal and out the negative terminal.
C—cycles back and forth with the number of cycles per second being controlled by the rotational speed of the alternator.

There are two ways of thinking about the direction of "current flow" in an electrical circuit. The actual movement of electrons, which are negative charges of electricity, is from negative to positive. Some references call this electron flow. An imaginary flow that follows the arrows in semiconductor symbols, and travels from positive to negative is called conventional current. The reference from which this question was taken considers current flow to be from negative to positive.

When charging a battery, current enters the negative terminal and leaves from the positive terminal.

8354. Aircraft that operate more than one generator connected to a common electrical system must be provided with

A—automatic generator switches that operate to isolate any generator whose output is less than 80 percent of its share of the load.
B—an automatic device that will isolate nonessential loads from the system if one of the generators fails.
C—individual generator switches that can be operated from the cockpit during flight.

Generator switches are required to allow the individual generators to be put on the line or taken off at the discretion of the pilot or flight engineer.

8355. The most effective method of regulating aircraft direct current generator output is to vary, according to the load requirements, the

A—strength of the stationary field.
B—generator speed.
C—number of rotating armature loops in use.

The output voltage of a DC generator is controlled by varying the strength of the generator stationary field.

The voltage regulator controls the amount of current flowing in the field coils.

8356. Electric motors are often classified according to the method of connecting the field coils and armature. Aircraft engine starter motors are generally of which type?

A—Compound.
B—Series.
C—Shunt (parallel).

Aircraft starters normally use series-wound DC motors because of their high starting torque.

Answers

8350 [B] (J01) AMT-G	8351 [C] (J01) AMT-P	8352 [C] (J01) AMT-G	8353 [A] (J01) AMT-G
8354 [C] (J01) 14 CFR 25.1351	8355 [A] (J01) AMT-G	8356 [B] (J01) AMT-G	

8357. As the generator load is increased (within its rated capacity), the voltage will

A—decrease and the amperage output will increase.
B—remain constant and the amperage output will increase.
C—remain constant and the amperage output will decrease.

The voltage regulator will hold the generator output voltage constant, but the amperage output (the current produced by the generator) will increase as the generator load is increased.

8358. As the flux density in the field of a dc generator increases and the current flow to the system increases, the

A—generator voltage decreases.
B—generator amperage decreases.
C—force required to turn the generator increases.

A generator is a device that converts mechanical energy from the engine into electrical energy to supply the electrical load.

When the load current flow in the system increases and the flux density in the field of the generator increases to produce more current, the force needed to turn the generator also increases.

8359. What is the purpose of a reverse-current cutout relay?

A—It eliminates the possibility of reversed polarity of the generator output current.
B—It prevents fluctuations of generator voltage.
C—It opens the main generator circuit whenever the generator voltage drops below the battery voltage.

The reverse-current cutout relay connected to a DC generator, opens the main generator circuit when the generator voltage drops below that of the battery.

8360. Generator voltage will not build up when the field is flashed and solder is found on the brush cover plate. These are most likely indications of

A—an open armature.
B—excessive brush arcing.
C—armature shaft bearings overheating.

If a generator does not produce voltage after its field is flashed, and molten solder is found on the brush cover plate, the armature is open. The molten solder is an indication of generator overheating.

8361. Why is it unnecessary to flash the field of the exciter on a brushless alternator?

A—The exciter is constantly charged by battery voltage.
B—Brushless alternators do not have exciters.
C—Permanent magnets are installed in the main field poles.

Permanent magnets are built into the field poles of a brushless alternator to provide enough magnetic flux to start the alternator producing electricity.

Because of these permanent magnets, there is no need to flash the field to put residual magnetism into the field frame.

8362. One way that the automatic ignition relight systems are activated on gas turbine engines is by a

A—drop in compressor discharge pressure.
B—sensing switch located in the tailpipe.
C—drop in fuel flow.

A drop in compressor discharge pressure is one of the first indications of a loss in engine power.

Some gas turbine engines have a pressure sensor that detects a drop in compressor discharge pressure and turns on the automatic ignition relight system.

8363. How are the rotor windings of an aircraft alternator usually excited?

A—By a constant ac voltage from the battery.
B—By a constant ac voltage.
C—By a variable direct current.

Most alternator rotors are excited by direct current from the aircraft battery, through the voltage regulator.

When the alternator load increases and the output voltage drops, the voltage regulator supplies more current to the alternator rotor coil.

8364. What precaution is usually taken to prevent electrolyte from freezing in a lead acid battery?

A—Place the aircraft in a hangar.
B—Remove the battery and keep it under constant charge.
C—Keep the battery fully charged.

When a lead-acid battery is fully charged, much of the water in the electrolyte has been replaced with sulfuric acid.

The freezing temperature of the electrolyte of a fully charged lead-acid battery is much lower than the freezing temperature of the electrolyte in a discharged battery.

Answers

8357 [B] (J01) AMT-G	8358 [C] (J01) AMT-SYS	8359 [C] (J01) AMT-G	8360 [A] (J01) AMT-G
8361 [C] (J01) AMT-G	8362 [A] (J01) AMT-P	8363 [C] (J01) AMT-G	8364 [C] (J01) AMT-G

8365. What is the ampere-hour rating of a storage battery that is designed to deliver 45 amperes for 2.5 hours?

A—112.5 ampere-hour.
B—90.0 ampere-hour.
C—45.0 ampere-hour.

The ampere-hour capacity of a battery is found by multiplying the amount of current the battery will supply by the number of hours it will supply this flow.

In this example, the battery that will supply 45 amperes for 2.5 hours has an ampere-hour capacity of 112.5 ampere-hours.

8366. How many hours will a 140 ampere-hour battery deliver 15 amperes?

A—1.40 hours.
B—9.33 hours.
C—14.0 hours.

The ampere-hour capacity of a battery is found by multiplying the amount of current the battery will supply by the number of hours it will supply this flow.

In this example, a 140-ampere-hour battery will supply 15 amperes of current for 9.33 hours.

8367. What is a basic advantage of using ac for electrical power for a large aircraft?

A—AC systems operate at higher voltage than dc systems and therefore use less current and can use smaller and lighter weight wiring.
B—AC systems operate at lower voltage than dc systems and therefore use less current and can use smaller and lighter weight wiring.
C—AC systems operate at higher voltage than dc systems and therefore use more current and can use smaller and lighter weight wiring.

An advantage of alternating current over direct current is that its voltage and current can be easily stepped up or down.

Large aircraft have applications for large amounts of electrical power. This may be supplied at a high voltage with low current. The low current allows the use of small gauge wires which saves weight.

8368. What are two types of ac motors that are used to produce a relatively high torque?

A—Shaded pole and shunt field.
B—Shunt field and single phase.
C—Three-phase induction and capacitor start.

The two types of AC motors that have the highest starting torque are the three-phase induction motors and single-phase induction motors with capacitor start.

8369. (1) Alternators are rated in volt-amps, which is a measure of the apparent power being produced by the generator.

(2) Alternating current has the advantage over direct current in that its voltage and current can easily be stepped up or down.

Regarding the above statements,

A—only No. 1 is true.
B—only No. 2 is true.
C—both No. 1 and No. 2 are true.

Both statements are true. AC generators are rated, not in watts, but in volt-amps, or kilovolt-amps, which is a measure of the apparent power produced by the generator.

Alternating current has the advantage over direct current in that its voltage and current can easily be stepped up or down.

8370. What is the frequency of most aircraft alternating current?

A—115 Hertz.
B—60 Hertz.
C—400 Hertz.

Almost all aircraft electrical systems use 400-hertz alternating current.

8371. The reason for flashing the field in a generator is to

A—restore correct polarity and/or residual magnetism to the field poles.
B—increase generator capacity.
C—remove excessive deposits.

A generator field is flashed by passing direct current from a battery through the field coils in the direction it normally flows.

This current restores the residual magnetism of the correct polarity to the field frame and allows the generator to begin producing current as soon as it begins to turn.

8372. The part of a dc alternator power system that prevents reverse flow of current from the battery to the alternator is the

A—reverse current relay.
B—voltage regulator.
C—rectifier.

Rectifiers in the form of semiconductor diodes prevent current from the battery flowing into the alternator.

Answers

8365 [A] (J01) AMT-G	8366 [B] (J01) AMT-G	8367 [A] (J01) AMT-G	8368 [C] (J01) AMT-G
8369 [C] (J01) AMT-G	8370 [C] (J01) AMT-G	8371 [A] (J01) AMT-P	8372 [C] (J01) AMT-P

8373. The generating system of an aircraft charges the battery by using

A—constant current and varying voltage.
B—constant voltage and varying current.
C—constant voltage and constant current.

The DC electrical system of an aircraft keeps the battery charged by supplying it with a constant voltage.

When the battery voltage is low, the generator supplies a large amount of charging current, but as the battery voltage rises, the charging current decreases

8374. The constant current method of charging a ni-cad battery

A—will bring it up to fully charged in the shortest amount of time.
B—will lead to cell imbalance over a period of time.
C—is the method most effective in maintaining cell balance.

The constant-current method of battery charging is the preferred method for charging nickel-cadmium batteries. It is slower than the constant-voltage method, but it is the most effective in maintaining cell balance and capacity.

8375. (Refer to Figure 4.) The following data concerning the installation of an electrical unit is known: current requirements for continuous operation—11 amperes; measured cable length—45 feet; system voltage—28 volts (do not exceed 1 volt drop); cable in conduit and bundles. What is the minimum size copper electrical cable that may be selected?

A—No. 10.
B—No. 12.
C—No. 14.

Draw a diagonal line downward and to the left, parallel to the 10-amp line, 1/5 of the way between the 10-amp and 15-amp diagonals.

Draw a line horizontally to the right from the 45-foot mark in the right-hand column. This is the column for a one-volt drop.

These two lines intersect between the vertical lines for a 12-gauge and a 14-gauge wire. Always select the larger wire.

A 12-gauge copper wire will carry 11 amperes for 45 feet without exceeding a voltage drop of one volt.

The intersection of these two lines is above curve 1 of the chart and this indicates that this load may be carried with the wire in a bundle or conduit.

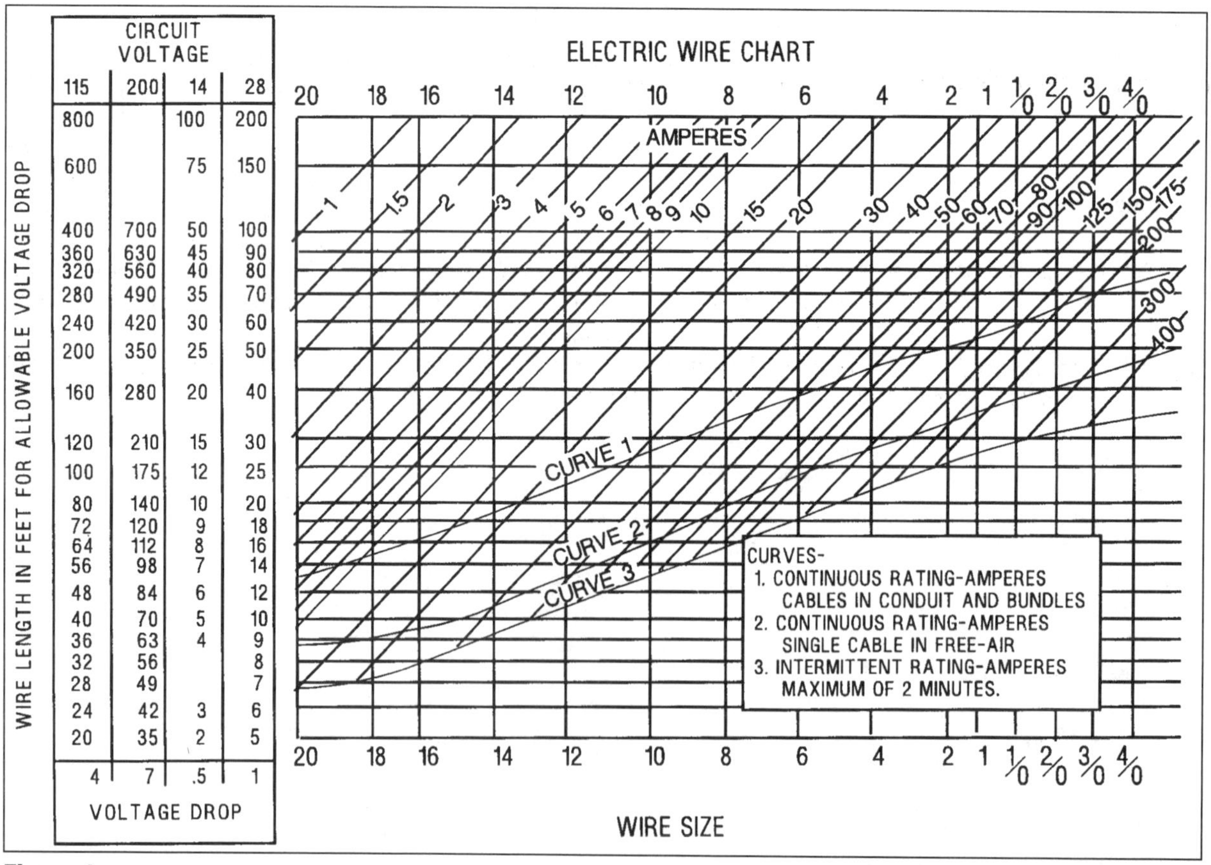

Figure 4

Answers
8373 [B] (J01) AMT-G
8374 [C] (J01) AMT-G
8375 [B] (J02) AC 43.13-1B

8376. Which of the following aircraft circuits does NOT contain a fuse/circuit breaker?

A—Generator circuit.
B—Air-conditioning circuit.
C—Starter circuit.

An aircraft starter circuit does not contain a fuse.

The extremely high current drawn in the locked-rotor condition (the condition when the current begins to flow) of an aircraft starter would make a fuse in this type of circuit impractical.

8377. The maximum number of terminals that may be connected to any one terminal stud in an aircraft electrical system is

A—two.
B—three.
C—four.

No more than four terminals should be connected to any one terminal stud.

If it is necessary to connect more than four wires to a single point, use two or more adjacent studs and mount a small metal bus strap across them.

In all cases, the current must be carried by the terminal contact surface and not by the stud itself.

8378. What is the maximum number of bonding jumper wires that may be attached to one terminal grounded to a flat surface?

A—Two.
B—Three.
C—Four.

It is a good practice to limit to four, the number of bonding jumpers attached to one terminal that is grounded to a flat surface.

In this way, the terminals may be spread out and good contact between the terminals is ensured.

8379. As a general rule, starter brushes are replaced when they are approximately

A—one-half their original length.
B—one-third their original length.
C—two-thirds their original length.

Generally, starter brushes are replaced when they are worn to about one-half their original length.

This wear is measured by comparing the worn brush with a new brush.

8380. When installing an electrical switch, under which of the following conditions should the switch be derated from its nominal current rating?

A—Conductive circuits.
B—Capacitive circuits.
C—Direct-current motor circuits.

An electrical switch must be derated from its normal current rating when it is used with DC motors because of the high inrush of current.

This high rate of current flow drops as soon as the motor starts turning and producing back EMF.

8381. The resistance of the current return path through the aircraft is always considered negligible, provided the

A—voltage drop across the circuit is checked.
B—generator is properly grounded.
C—structure is adequately bonded.

The resistance of the current return path through an aircraft structure is considered to be negligible as long as the structure is adequately bonded (connected together electrically with low-resistance bonding straps or braids).

A resistance of 3 milliohms (0.003 ohm) between the ground and the generator, or battery, is considered to be satisfactory.

There must be no measurable voltage drop across any of the bonding straps.

Answers

8376 [C] (J02) AMT-P	8377 [C] (J02) AC 43.13-1B	8378 [C] (J02) AC 43.13-1B	8379 [A] (J02) AMT-P
8380 [C] (J02) AMT-P	8381 [C] (J02) AMT-G		

8382. In order to reduce the possibility of ground shorting the circuits when the connectors are separated for maintenance, the AN and MS electrical connectors should be installed with the

A—socket section on the ground side of the electrical circuit.
B—pin section on the ground side of the electrical circuit.
C—pin section on the positive side of the electrical circuit.

When installing AN or MS electrical connectors in a circuit, the half of the connector containing the sockets should be installed on the "hot" side of the circuit, and the half containing the pins on the ground side.

8383. When does current flow through the coil of a solenoid-operated electrical switch?

A—Continually, as long as the aircraft's electrical system master switch is on.
B—Continually, as long as the control circuit is complete.
C—Only until the movable points contact the stationary points.

Current flows through the coil of a solenoid-operated switch as long as the control circuit is complete.

When the control circuit is open, current stops flowing through the coil and a spring separates the contacts in the solenoid-operated switch.

8384. It is necessary to determine that the electrical load limit of a 28-volt, 75-amp generator, installed in a particular aircraft, has not been exceeded. By making a ground check, it is determined that the battery furnished 57 amperes to the system when all equipment that can continuously draw electrical power in flight is turned on. This type of load determination

A—can be made, but the load will exceed the generator load limit.
B—can be made, and the load will be within the generator load limit.
C—cannot be made on direct current electrical systems.

When this type of load determination is made, we see that the battery furnishes 57 amps to the system. It does this at the 24 volts of the battery. At the 28 volts put out by the generator, 66.5 amps will flow.

If this load can be monitored, it is within the generator load limit. However, if there is no way to monitor the amount of load on the system, the load should be restricted to 60 amps, which is 80% of the generator output.

8385. What type of lubricant may be used to aid in pulling electrical wires or cables through conduits?

A—Silicone grease.
B—Soapstone talc.
C—Rubber lubricant.

Soapstone talc, such as tire talcum, may be used to lubricate the inside of an electrical conduit or vinyl tubing to aid electrical wires being pulled through.

No other type of lubricant should be used.

8386. Which of the following is regulated in a generator to control its voltage output?

A—Speed of the armature.
B—Number of windings in the armature.
C—The strength of the field.

The output voltage of a generator is regulated by controlling the amount of current allowed to flow in its field coils.

8387. Bonding jumpers should be designed and installed in such a manner that they

A—are not subjected to flexing by relative motion of airframe or engine components.
B—provide a low electrical resistance in the ground circuit.
C—prevent buildup of a static electrical charge between the airframe and the surrounding atmosphere.

Bonding jumpers should be as short as practicable and installed in such a manner that the resistance of each connection does not exceed 0.003 ohm. This ensures a low electrical resistance in the ground circuit.

Answers

8382 [B] (J02) AMT-G
8383 [B] (J02) AMT-G
8384 [B] (J02) AMT-P
8385 [B] (J02) AMT-P
8386 [C] (J02) AMT-G
8387 [B] (J02) AC 43.13-1B

8388. When the starter switch to the aircraft gas turbine engine starter-generator is energized and the engine fails to rotate, one of the probable causes would be the

A—power lever switch is defective.
B—undercurrent solenoid contacts are defective.
C—starter solenoid is defective.

When the starter-generator circuit shown in Figure 5, Page 92 (also in Powerplant Test Figures booklet) is energized, starting current must flow from the bus through the starter solenoid contacts and the coil of the undercurrent relay to the "C" terminal of the starter-generator. If the starter solenoid is defective, it will not allow current to reach the starter windings.

8389. Arcing at the brushes and burning of the commutator of a motor may be caused by

A—weak brush springs.
B—excessive brush spring tension.
C—low mica.

Weak brush springs allow the brushes of a motor to bounce and cause arcing between the brushes and the commutator.

8390. The maximum allowable voltage drop between the generator and the bus bar is

A—1 percent of the regulated voltage.
B—2 percent of the regulated voltage.
C—less than the voltage drop permitted between the battery and the bus bar.

The maximum allowable voltage drop between the generator and the bus bar is 2% of the regulated voltage of the generator.

8391. ON-OFF two position engine electrical switches should be installed

A—so that the toggle will move in the same direction as the desired motion of the unit controlled.
B—under a guard.
C—so the ON position is reached by a forward or upward motion.

The standard installation practice for engine controls (and this includes electrical switches) is for a forward or upward movement of the control to cause an "ON" or an "INCREASE" condition.

8392. When selecting an electrical switch for installation in an aircraft circuit utilizing a direct current motor,

A—a switch designed for dc should be chosen.
B—a derating factor should be applied.
C—only switches with screw-type terminal connections should be used.

DC electric motors have a high inrush of current when the switch is first closed. This is because the current-limiting counter electromotive force is not produced until the motor is turning.

Switches used in an electric motor circuit must be derated by a factor of three for a 24-volt motor and by a factor of two for a 12-volt motor.

8393. When installing electrical wiring parallel to a fuel line, the wiring should be

A—in metal conduit.
B—in a non-conductive fire-resistant sleeve.
C—above the fuel line.

Anytime an electrical wire bundle is installed in an aircraft in such a way that it is installed parallel with a fuel line, the wire bundle must be above the fuel line.

If the fuel line should leak, it must not leak into the wire bundle.

Answers

8388 [C] (J02) AMT-P
8389 [A] (J02) AMT-P
8390 [B] (J02) AC 43.13-1B
8391 [C] (J02) AC 43.13-1B
8392 [B] (J02) AMT-P
8393 [C] (J02) AC 43.13-1B

8394. (Refer to Figure 4.) In a 28-volt system, what is the maximum continuous current that can be carried by a single No. 10 copper wire 25 feet long, routed in free air?

A—20 amperes.
B—35 amperes.
C—28 amperes.

Follow the 25-foot horizontal line in the one volt drop (28-volt) column to the right until it intersects the vertical line for the 10-gauge wire.

This intersection occurs approximately midway between the diagonal lines for 30 amps and 40 amps. This means that a 10-gauge wire can carry 35 amps for 25 feet and not have more than a one-volt drop in a 28-volt circuit.

The intersection is below curve 1, which means that this wire could not carry this much current if it were routed in a bundle, but it is well above curve 2, so it can carry this much current as a single wire routed in free air.

8395. What speed must an eight-pole ac generator turn to produce 400-Hertz ac?

A—400 RPM.
B—1,200 RPM.
C—6,000 RPM.

To find the speed an eight-pole generator must turn to produce 400-hertz alternating current, multiply the frequency by 120. Then divide this by the number of poles.

400 × 120 ÷ 8 = 6,000 RPM

8396. How many basic types of circuit breakers are used in powerplant installation electrical systems?

A—Two.
B—Three.
C—Four.

There are three basic types of circuit breakers used in aircraft electrical systems: push-to-reset, push-pull, and toggle-type.

8397. Which Federal Aviation Regulation specifies that each resettable circuit protective device requires a manual operation to restore service after the device has interrupted the circuit?

A—14 CFR Part 23.
B—14 CFR Part 43.
C—14 CFR Part 91.

This requirement regarding circuit protective devices is found in 14 CFR §23.1357(c)(1). "A manual operation is required to restore service after tripping;..."

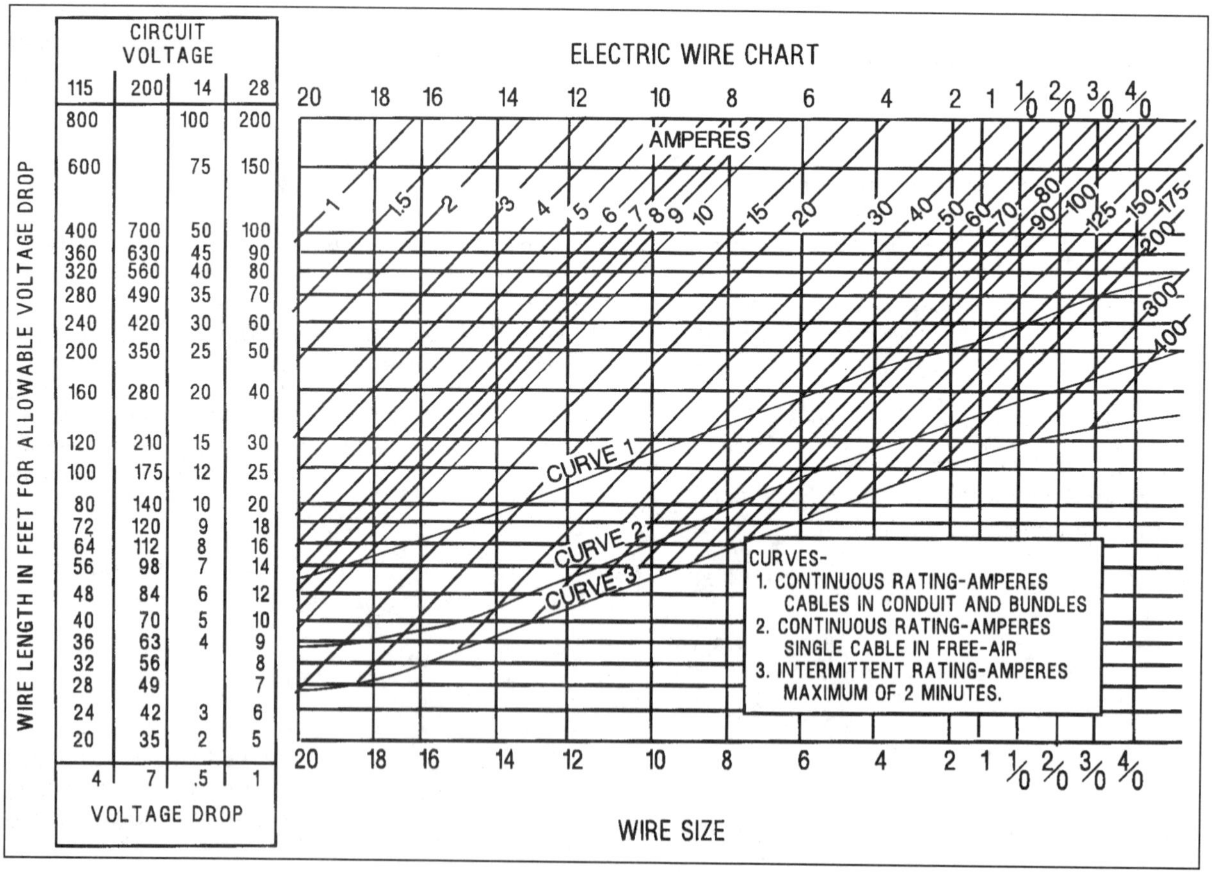

Figure 4

Answers

8394 [B] (J02) AC 43.13-1B
8395 [C] (J02) AMT-G
8396 [B] (J02) AMT-G
8397 [A] (J02) AMT-P

8398. Which Federal Aviation Regulation requirement prevents the use of automatic reset circuit breakers?

A—14 CFR Part 21.
B—14 CFR Part 23.
C—14 CFR Part 91.

14 CFR §23.1357(c)(1) does not allow the use of automatic-reset circuit breakers because it states: "A manual operation is required to restore service after tripping."

8399. The time/current capacities of a circuit breaker or fuse must be

A—above those of the associated conductor.
B—equal to those of the associated conductor.
C—below those of the associated conductor.

Fuse or circuit breakers are installed in a circuit primarily to protect the wiring. In order to do this, the time/current capacities of the circuit protection device must be below that of the associated conductor.

8400. (1) Most modern aircraft use circuit breakers rather than fuses to protect their electrical circuits.

(2) Federal Aviation Regulations Part 23 requires that all electrical circuits incorporate some form of circuit protective device.

Regarding the above statements,

A—only No. 1 is true.
B—only No. 2 is true.
C—both No. 1 and No. 2 are true.

Statement (1) is true. Most modern aircraft use circuit breakers rather than fuses to protect electrical circuits.

Statement (2) is not true. 14 CFR §23.1357 states that, "Protective devices such as fuses or circuit breakers, must be installed in all electrical circuits other than the main circuits of starter motors, and circuits in which no hazard is presented by their omission."

8401. Electrical switches are rated according to the

A—voltage and the current they can control.
B—resistance rating of the switch and the wiring.
C—resistance and the temperature rating.

Electrical switches are rated according to both the voltage and the current they can control.

8402. Electrical circuit protection devices are installed primarily to protect the

A—switches.
B—units.
C—wiring.

Electrical-circuit protection devices, such as fuses and circuit breakers, are installed primarily to protect the wiring. The rating of the device is based on the amount of current the wire can carry without overheating its insulation.

8403. (1) Electrical circuit protection devices are rated based on the amount of current that can be carried without overheating the wiring insulation.

(2) A "trip-free" circuit breaker makes it impossible to manually hold the circuit closed when excessive current is flowing.

Regarding the above statements,

A—only No. 1 is true.
B—only No. 2 is true.
C—both No. 1 and No. 2 are true.

Both statements are true.

The rating of electrical circuit protection devices is based on the amount of current that can be carried without overheating the wiring insulation.

A "trip-free" circuit breaker will keep a circuit open in the presence of a fault, regardless of the position of the operating handle.

8404. Which of the following Federal Aviation Regulations require that all aircraft using fuses as the circuit protective devices carry "one spare set of fuses, or three spare fuses of each kind required"?

A—14 CFR Part 23.
B—14 CFR Part 43.
C—14 CFR Part 91.

14 CFR §91.205(c)(6) requires that all aircraft operating under visual flight rules (night) carry one spare set of fuses or three spare fuses of each kind required. This same requirement applies to aircraft operating under instrument flight rules.

8405. What is the smallest terminal stud allowed for aircraft electrical power systems?

A—No. 6.
B—No. 8.
C—No. 10.

A number 10 stud is the smallest recommended for use in the electrical power system of an aircraft. Smaller studs, however, are used in some operational systems in the aircraft.

Answers

8398 [B] (J02) AMT-P	8399 [C] (J02) AMT-P	8400 [A] (J02) 14 CFR 23.1357	8401 [A] (J02) AMT-P
8402 [C] (J02) AMT-P	8403 [C] (J02) AMT-G	8404 [C] (J02) AMT-P	8405 [C] (J02) AMT-P

8406. A typical barrier type aircraft terminal strip is made of

A—paper-base phenolic compound.
B—polyester resin and graphite compound.
C—layered aluminum impregnated with compound.

Many of the terminal strips used in aircraft electrical systems are of the barrier type and are made of a strong, paper-base phenolic compound.

8407. A term commonly used when two or more electrical terminals are installed on a single lug of a terminal strip is

A—strapping.
B—stepping.
C—stacking.

Stacking is a term used to describe the placement of two or more terminals on a single lug or stud of an electrical terminal strip.

8408. (1) Electrical wires larger than 10 gauge use uninsulated terminals.

(2) Electrical wires smaller than 10 gauge use uninsulated terminals.

Regarding the above statements,

A—only No. 1 is true.
B—only No. 2 is true.
C—neither No. 1 nor No. 2 is true.

Statement (1) is true. Electrical wires larger than 10-gauge, use uninsulated terminals. After the terminal is swaged onto the wire, it is insulated with heat-shrink tubing or vinyl tubing.

Statement (2) is not true. Electrical wires smaller than 10-gauge, normally use preinsulated terminals. The color of the insulation shows the size of wire the terminal fits.

8409. Aircraft electrical wire size is measured according to the

A—Military Specification system.
B—American Wire Gauge system.
C—Technical Standard Order system.

Aircraft electrical wire is measured according to the American Wire Gauge (AWG) system.

8410. Aircraft copper electrical wire is coated with tin, silver, or nickel in order to

A—improve conductivity.
B—add strength.
C—prevent oxidization.

Aircraft copper wire is covered with a very thin coating of tin, silver, or nickel to prevent oxidation.

8411. What will be the result of operating an engine in extremely high temperatures using a lubricant recommended by the manufacturer for a much lower temperature?

A—The oil pressure will be higher than normal.
B—The oil temperature and oil pressure will be higher than normal.
C—The oil pressure will be lower than normal.

If an oil recommended for low-temperature operation is used in an engine when it is operating in high-temperature conditions, the oil will have a lower viscosity than should be used in the engine. Therefore, the oil pressure will be lower than recommended.

8412. (1) Gas turbine and reciprocating engine oils can be mixed or used interchangeably.

(2) Most gas turbine engine oils are synthetic.

Regarding the above statements,

A—only No. 2 is true.
B—both No. 1 and No. 2 are true.
C—neither No. 1 nor No. 2 is true.

Statement (1) is not true. The lubricating requirements of a turbine engine are quite different from those of a reciprocating engine. Therefore, the types of oil used in the two engines are different.

Statement (2) is true because most of the lubricating oils used in turbine engines have a synthetic base.

8413. An oil separator is generally associated with which of the following?

A—Engine-driven oil pressure pump.
B—Engine-driven vacuum pump.
C—Cuno oil filter.

An oil separator is used with a "wet"-type engine-driven vacuum pump.

Wet-type vacuum pumps are lubricated by oil flowing through the pump and it is discharged with the air that leaves the pump.

An oil separator in the discharge line separates the oil from the air. The air is discharged overboard or used in the deicer boots and the oil is returned to the engine crankcase.

Answers

8406 [A] (J02) AMT-P	8407 [C] (J02) AMT-P	8408 [A] (J02) AMT-P	8409 [B] (J02) AMT-G
8410 [C] (J02) AMT-P	8411 [C] (K01) AMT-P	8412 [A] (K01) AMT-P	8413 [B] (K01) AMT-SYS

8414. The time in seconds required for exactly 60 cubic centimeters of oil to flow through an accurately calibrated orifice at a specific temperature is recorded as a measurement of the oil's

A—flash point.
B—specific gravity.
C—viscosity.

The method of determining the viscosity of an oil described here is the procedure used to find the Seconds Saybolt Universal (SSU) viscosity of the oil.

The SSU viscosity is usually converted into SAE viscosity which is more familiar to A&P mechanics.

8415. Upon what quality or characteristic of a lubricating oil is its viscosity index based?

A—Its resistance to flow at a standard temperature as compared to high grade paraffin-base oil at the same temperature.
B—Its rate of change in viscosity with temperature change.
C—Its rate of flow through an orifice at a standard temperature.

The viscosity index of an oil is a measure of the rate of change of the viscosity of the oil with a change in its temperature.

8416. Lubricating oils with high viscosity index ratings are oils

A—in which the viscosity does not vary much with temperature change.
B—in which the viscosity varies considerably with temperature change.
C—which have high SAE numbers.

The viscosity index of an oil is a measure of the rate of change of the viscosity with a change in its temperature.

The higher the viscosity index of an oil, the less the viscosity changes as its temperature changes.

8417. Compared to reciprocating engine oils, the types of oils used in turbine engines

A—are required to carry and disperse a higher level of combustion by-products.
B—may permit a somewhat higher level of carbon formation in the engine.
C—have less tendency to produce lacquer or coke.

Synthetic oil used in turbine engines has two advantages over petroleum oil. It has less tendency to deposit lacquer and coke and it has less tendency to evaporate at high temperature.

8418. The oil used in reciprocating engines has a relatively high viscosity due to

A—the reduced ability of thin oils to maintain adequate film strength at altitude (reduced atmospheric pressure).
B—the relatively high rotational speeds.
C—large clearances and high operating temperatures.

The oil used in reciprocating engines has a relatively high viscosity because of:

1. *Large engine operating clearances due to the relatively large size of the moving parts, the different materials used and the different rates of expansion of the various materials*
2. *High operating temperatures*
3. *High bearing pressures*

8419. If all other requirements can be met, what type of oil should be used to achieve theoretically perfect engine lubrication?

A—The thinnest oil that will stay in place and maintain a reasonable film strength.
B—An oil that combines high viscosity and low demulsibility.
C—An oil that combines a low viscosity index and a high neutralization number.

An ideal engine oil is one with a low viscosity (it is thin and pours easily). The low viscosity allows it to circulate easily within the engine.

It must also maintain a reasonable film strength so the lubricating film will not break down at the high operating pressures and temperatures encountered within the engine.

8420. In addition to lubricating (reducing friction between moving parts), engine oil performs what functions?

1. Cools.
2. Seals.
3. Cleans.
4. Prevents corrosion.
5. Cushions impact (shock) loads.

A—1, 2, 3, 4.
B—1, 2, 3, 4, 5.
C—1, 3, 4.

In addition to reducing friction between moving parts, an engine lubricant accomplishes all five of the items listed in the alternatives for this question.

It cools, seals, cleans, prevents corrosion, and serves as a cushion between parts where impact loads are involved.

Answers

8414 [C] (K01) AMT-P	8415 [B] (K01) AMT-P	8416 [A] (K01) AMT-P	8417 [C] (K01) AMT-P
8418 [C] (K01) AMT-P	8419 [A] (K01) AMT-P	8420 [B] (K01) AMT-P	

8421. Which of these characteristics is desirable in turbine engine oil?

A—Low flash point.
B—High flash point.
C—High volatility.

An engine lubricating oil should have a high flash point so that the vapors it gives off can withstand the high temperatures encountered within the engine.

The other alternatives listed with this question are characteristics an oil should not have.

8422. The viscosity of a liquid is a measure of its

A—resistance to flow.
B—rate of change of internal friction with change in temperature.
C—weight, or density.

The viscosity of a liquid is a measure of its ability to flow at a specific temperature. This can be thought of as its resistance to flow.

8423. What type of oil system is usually found on turbine engines?

A—Dry sump, pressure, and spray.
B—Dry sump, dip, and splash.
C—Wet sump, spray, and splash.

Most turbojet engines use a dry-sump lubrication system.

A pressure pump directs the oil to the oil jets or nozzles located in the pressure lines adjacent to or within the bearing compartments and the rotor shaft couplings.

The oil is delivered to the bearings in the form of an atomized spray.

8424. Which of the following factors helps determine the proper grade of oil to use in a particular engine?

A—Adequate lubrication in various attitudes of flight.
B—Positive introduction of oil to the bearings.
C—Operating speeds of bearings.

The operating load, the rotational speed (the operating speed of the bearing) and the operating temperatures are the most important considerations in determining the proper grade of oil to use in an aircraft engine.

8425. Specific gravity is a comparison of the weight of a substance to the weight of an equal volume of

A—oil at a specific temperature.
B—distilled water at a specific temperature.
C—mercury at a specific temperature.

Specific gravity is the ratio of the weight of a definite volume of the material being measured to an equal volume of pure water (distilled water) at its maximum density (at 4°C).

8426. Which of the following has the greatest effect on the viscosity of lubricating oil?

A—Temperature.
B—Pressure.
C—Volatility.

Temperature has a greater effect than any of the other choices given here on the viscosity of a lubricating oil.

8427. What advantage do mineral base lubricants have over vegetable oil base lubricants when used in aircraft engines?

A—Cooling ability.
B—Chemical stability.
C—Friction resistance.

Mineral-based lubricating oils have a much greater chemical stability than do vegetable-based lubricants.

8428. The recommended aircraft engine lubricants are

A—animal, mineral, or synthetic based.
B—mineral or synthetic based.
C—vegetable, mineral, or synthetic based.

The most satisfactory lubricants for aircraft engines have either a mineral base or a synthetic base.

8429. High tooth pressures and high rubbing velocities, such as occur with spur-type gears, require the use of

A—an EP lubricant.
B—straight mineral oil.
C—metallic ash detergent oil.

Some spur gears, and hypoid-type gearing having high tooth pressures and high rubbing velocities, require the use of extreme-pressure (EP) lubricants.

8430. Manufacturers normally require turbine engine oil servicing within a short time after engine shutdown primarily to

A—prevent overservicing.
B—help dilute and neutralize any contaminants that may already be present in the engine's oil system.
C—provide a better indication of any oil leaks in the system.

Answers

8421 [B] (K01) AMT-P
8425 [B] (K01) AMT-P
8422 [A] (K01) AMT-P
8426 [A] (K01) AMT-P
8423 [A] (K01) AMT-P
8427 [B] (K01) AMT-P
8429 [A] (K01) AMT-P
8424 [C] (K01) AMT-P
8428 [B] (K01) AMT-P
8430 [A] (K01) AMT-P

Turbine engine manufacturers normally recommend engine servicing to be accomplished as soon as practical after engine shutdown to prevent overservicing.

Overservicing may occur because, in some engines, the oil from the storage tank seeps into the lower portions of the engine after the engine sits for some time without being operated.

8431. What type of oil do most engine manufacturers recommend for new reciprocating engine break-in?

A—Ashless-dispersant oil.
B—Straight mineral oil.
C—Semi-synthetic oil.

Some engine manufacturers recommend that new or freshly overhauled engines be operated with straight mineral oil for the first fifty hours or at least until oil consumption stabilizes. After this, AD oil can be used.

8432. What type of oil do most engine manufacturers recommend after new reciprocating engine break-in?

A—Metallic-ash detergent oil.
B—Ashless-dispersant oil.
C—Straight mineral oil.

After a reciprocating engine is adequately broken in with straight mineral oil, it is recommended that ashless-dispersant (AD) oil be used.

AD oil has better lubricating characteristics than straight mineral oil and it is not so likely to form carbon deposits on the engine parts.

8433. The type of oil pumps most commonly used on turbine engines are classified as

A—positive displacement.
B—variable displacement.
C—constant speed.

The oil pumps used in turbojet engine lubrication systems are of the positive displacement type.

They may be either the vane type, spur gear type, or the gerotor type.

8434. As a general rule, the mixture setting on a reciprocating engine operating at or near takeoff power that provides the best cooling is

A—FULL RICH.
B—LEAN.
C—FULL LEAN.

The mixture control of the carburetor or fuel injection system installed on a reciprocating engine should be set in the FULL RICH position for takeoff and high power operation. This mixture provides more fuel than is needed for optimum performance, and the additional fuel is used for engine cooling.

8435. The engine oil temperature regulator is usually located between which of the following on a dry sump reciprocating engine?

A—The engine oil supply pump and the internal lubrication system.
B—The scavenger pump outlet and the oil storage tank.
C—The oil storage tank and the engine oil supply pump.

The oil-temperature regulator on a dry-sump reciprocating aircraft engine controls the temperature of the oil by directing it either through the core of the cooler or around the core.

The oil cooler is located between the scavenger-pump outlet and the storage tank.

8436. What will happen to the return oil if the oil line between the scavenger pump and the oil cooler separates?

A—Oil will accumulate in the engine.
B—The return oil will be pumped overboard.
C—The scavenger return line check valve will close and force the oil to bypass directly to the intake side of the pressure pump.

The scavenger pump pumps the oil from the engine through the cooler to the oil tank.

If the line between the scavenger pump and the oil cooler separates, the scavenger pump will pump all of the rcturn oil overboard.

8437. At cruise RPM, some oil will flow through the relief valve of a gear-type engine oil pump. This is normal as the relief valve is set at a pressure which is

A—lower than the pump inlet pressure.
B—lower than the pressure pump capabilities.
C—higher than pressure pump capabilities.

At cruise RPM, the engine oil pressure pump produces more flow at the pressure at which the oil-pressure relief valve is set than is needed to lubricate the engine.

Some of the oil will bypass back into the inlet side of the pump through the oil pressure relief valve.

Answers

8431 [B] (K01) AMT-P	8432 [B] (K01) AMT-P	8433 [A] (K02) AMT-P	8434 [A] (K02) AMT-P
8435 [B] (K02) AMT-P	8436 [B] (K02) AMT-P	8437 [B] (K02) AMT-P	

8438. (1) Fuel may be used to cool oil in gas turbine engines.

(2) Ram air may be used to cool oil in gas turbine engines.

Regarding the above statements,

A—only No. 1 is true.
B—only No. 2 is true.
C—both No. 1 and No. 2 are true.

Statement (1) is true. Fuel may be used to cool the engine oil in an oil-to-fuel heat exchanger.

Statement (2) is also true. Ram air may be used to cool the engine oil in an air-to-oil heat exchanger.

8439. In a reciprocating engine oil system, the temperature bulb senses oil temperature

A—at a point after the oil has passed through the oil cooler.
B—while the oil is in the hottest area of the engine.
C—immediately before the oil enters the oil cooler.

In a reciprocating engine lubrication system, the oil temperature bulb senses the oil temperature at a point after the oil has passed through the oil cooler. The temperature shown in the cockpit is the temperature of the oil as it enters the engine.

8440. The oil dampened main bearing utilized in some turbine engines is used to

A—provide lubrication of bearings from the beginning of starting rotation until normal oil pressure is established.
B—provide an oil film between the outer race and the bearing housing in order to reduce vibration tendencies in the rotor system, and to allow for slight misalignment.
C—dampen surges in oil pressure to the bearings.

A main bearing oil damper compartment provides an oil film between the bearing outer race and the bearing housing to reduce vibration tendencies in the rotor system and to allow for a slight misalignment.

8441. What is the purpose of the last chance oil filters?

A—To prevent damage to the oil spray nozzle.
B—To filter the oil immediately before it enters the main bearings.
C—To assure a clean supply of oil to the lubrication system.

The last-chance oil filters are used inside a turbine engine to filter the oil immediately before it enters the main bearings.

8442. In a jet engine which uses a fuel-oil heat exchanger, the oil temperature is controlled by a thermostatic valve that regulates the flow of

A—fuel through the heat exchanger.
B—both fuel and oil through the heat exchanger.
C—oil through the heat exchanger.

The oil temperature in a turbojet-engine lubricating system is controlled by a thermostatic valve that regulates the flow of oil through the heat exchanger. The oil may bypass the heat exchanger if no cooling is needed.

8443. What prevents pressure within the lubricating oil tank from rising above or falling below ambient pressure (reciprocating engine)?

A—Oil tank check valve.
B—Oil pressure relief valve.
C—Oil tank vent.

The oil tank vent prevents the pressure in the oil tank used with a reciprocating engine rising above or falling below the ambient pressure.

8444. In an axial-flow turbine engine, compressor bleed air is sometimes used to aid in cooling the

A—fuel.
B—inlet guide vanes.
C—turbine, vanes, blades, and bearings.

Compressor bleed-air, taken from an axial-flow turbine engine, is passed through hollow guide vanes and turbine blades to help cool them. It is also used to cool the bearings.

8445. Oil picks up the most heat from which of the following turbine engine components?

A—Rotor coupling.
B—Compressor bearing.
C—Turbine bearing.

In a turbojet engine, the oil picks the most heat from the turbine bearings.

8446. Which of the following is a function of the fuel-oil heat exchanger on a turbojet engine?

A—Aerates the fuel.
B—Emulsifies the oil.
C—Increases fuel temperature.

The oil-to-fuel heat exchanger used with a turbojet engine, not only decreases the temperature of the oil, but also increases the temperature of the fuel to prevent fuel icing.

Answers

8438 [C] (K02) AMT-P
8442 [C] (K02) AMT-P
8439 [A] (K02) AMT-P
8443 [C] (K02) AMT-P
8440 [B] (K02) AMT-P
8444 [C] (K02) AMT-P
8441 [B] (K02) AMT-P
8445 [C] (K02) AMT-P
8446 [C] (K02) AMT-P

8447. According to Federal Aviation Regulations, oil tank fillers on turbine engines must be marked with the word

A—"oil" and the type and grade of oil specified by the manufacturer.
B—"oil" and tank capacity.
C—"oil."

Federal Aviation Regulation §33.71(c)(5) states that each turbine engine oil tank filler must be marked with the word "Oil." The capacity of the tank is not required.

8448. After making a welded repair to a pressurized-type turbine engine oil tank, the tank should be pressure checked to

A—not less than 5 PSI plus the maximum operating pressure of the tank.
B—not less than 5 PSI plus the average operating pressure of the tank.
C—5 PSI.

14 CFR §33.71 (c)(9) states that each pressurized turbine engine oil tank may not leak when subjected to maximum operating temperature and an internal pressure of not less than 5 psi plus the maximum operating pressure of the tank.

8449. Why are fixed orifice nozzles used in the lubrication system of gas turbine engines?

A—To provide a relatively constant oil flow to the main bearings at all engine speeds.
B—To keep back pressure on the oil pump, thus preventing an air lock.
C—To protect the oil seals by preventing excessive pressure from entering the bearing cavities.

Pressurized oil is distributed to the engine main bearings in a turbojet engine through fixed orifice nozzles that provide a relatively constant oil flow at all engine operating speeds.

8450. Possible failure related ferrous-metal particles in turbine engine oil cause an (electrical) indicating-type magnetic chip detector to indicate their presence by

A—disturbing the magnetic lines of flux around the detector tip.
B—bridging the gap between the detector center (positive) electrode and the ground electrode.
C—generating a small electric current that is caused by the particles being in contact with the dissimilar metal of the detector tip.

A magnetic chip detector warns of possible impending engine failure by indicating the presence of ferrous metal particles. If these particles are present in the oil, they will bridge the gap in the detector between the center electrode and ground electrode, completing an electrical circuit to illuminate a warning light in the cockpit.

8451. What would be the probable result if the oil system pressure relief valve should stick in the open position on a turbine engine?

A—Increased oil pressure.
B—Decreased oil temperature.
C—Insufficient lubrication.

If the oil-pressure relief valve in a turbojet engine stuck in the open position, oil would flow through it rather than going through the passages in the lubrication system.

This would result in insufficient lubrication of the engine.

8452. What is the primary purpose of the oil-to-fuel heat exchanger?

A—Cool the fuel.
B—Cool the oil.
C—De-aerate the oil.

The primary purpose of the oil-to-fuel heat exchanger is to remove heat from the engine lubricating oil (to cool the oil).

A secondary purpose of the oil-to-fuel heat exchanger is to raise the temperature of the fuel to prevent water in the fuel from precipitating out and freezing on the fuel filters.

8453. What unit in an aircraft engine lubrication system is adjusted to maintain the desired system pressure?

A—Oil pressure relief valve.
B—Oil viscosity valve.
C—Oil pump.

The oil-pressure relief valve in the lubrication system of an aircraft engine is adjusted to maintain the desired system pressure.

Pressure above that for which the relief valve is set, lifts the valve off its seat and bypasses the oil that caused the excess pressure back to the inlet side of the oil pump.

Answers

8447 [C] (K02) 14 CFR 33.71
8448 [A] (K02) 14 CFR 33.71
8449 [A] (K02) AMT-P
8450 [B] (K02) AMT-P
8451 [C] (K02) AMT-P
8452 [B] (K02) AMT-P
8453 [A] (K02) AMT-P

8454. Low oil pressure can be detrimental to the internal engine components. However, high oil pressure

A—should be limited to the engine manufacturer's recommendations.
B—has a negligible effect.
C—will not occur because of pressure losses around the bearings.

Oil pressure above that recommended by the engine manufacturer should be avoided because excessive oil pressure can damage oil coolers and burst oil lines. A burst oil line could cause a loss of all of the lubricating oil.

8455. What is the primary purpose of the oil breather pressurization system that is used on turbine engines?

A—Prevents foaming of the oil.
B—Allows aeration of the oil for better lubrication because of the air/oil mist.
C—Provides a proper oil spray pattern from the main bearing oil jets.

The primary purpose of the breather pressurization system in a turbine engine is to ensure a proper oil spray pattern from the main bearing oil jets and to furnish a pressure head to the scavenge system.

8456. The purpose of directing bleed air to the outer turbine case on some engines is to

A—provide optimum turbine blade tip clearance by controlling thermal expansion.
B—provide up to 100 percent kinetic energy extraction from the flowing gases.
C—allow operation in a thermal environment 600 to 800°F above the temperature limits of turbine blade and vane alloys.

The clearance between the tip of the turbine blades and the outer turbine case is critical to engine performance. As the engine gets hot, the case expands and this clearance increases.

This clearance must be kept at an absolute minimum, so some of the newer turbine engines duct compressor bleed air or air from the fan stream around the turbine case. The cools the case and causes it to shrink and decreases the gap between the blades and the case.

8457. Some larger reciprocating engines use a compensating oil pressure relief valve to

A—provide a high engine oil pressure when the oil is cold and automatically lower the oil pressure when the oil warms up.
B—compensate for changes in atmospheric pressure that accompany altitude changes.
C—automatically keep oil pressure nearly the same whether the oil is warm or cold.

Many large aircraft reciprocating engines have a compensating oil pressure relief valve that allows the oil pressure for cold oil to be considerably higher than it allows for warm oil. This higher pressure allows the thicker, higher viscosity oil to be forced through the engine bearings.

The plunger of the oil-pressure relief valve is held down by two springs when the oil is cold. However, when the oil warms up, a thermostatic valve opens and allows oil pressure to remove the force of one of the springs. For normal operation, only one spring holds the pressure relief valve on its seat.

8458. In order to relieve excessive pump pressure in an engine's internal oil system, most engines are equipped with a

A—vent.
B—bypass valve.
C—relief valve.

Almost all aircraft engines have a pressure pump whose output exceeds the demands of the engine.

A relief valve is incorporated in the pressure portion of the system to relieve excess pressure back to the inlet of the pump.

8459. What is the source of most of the heat that is absorbed by the lubricating oil in a reciprocating engine?

A—Crankshaft main bearings.
B—Exhaust valves.
C—Pistons and cylinder walls.

Engine lubricating oil absorbs as much heat as possible from all of the lubricated surfaces, but they absorb the most heat from the under side of the piston head and from the cylinder walls.

Answers
8454 [A] (K02) AMT-P 8455 [C] (K02) AMT-P 8456 [A] (K02) AMT-P 8457 [A] (K03) AMT-P
8458 [C] (K03) AMT-P 8459 [C] (K03) AMT-P

8460. How are the teeth of the gears in the accessory section of an engine normally lubricated?

A— By splashed or sprayed oil.
B— By submerging the load-bearing portions in oil.
C— By surrounding the load-bearing portions with baffles or housings within which oil pressure can be maintained.

The gear teeth in the accessory section of an engine are lubricated by oil that is sprayed from the accessory shaft bearings and splashed around inside the accessory case.

8461. What is the purpose of the check valve generally used in a dry sump lubrication system?

A— To prevent the scavenger pump from losing its prime.
B— To prevent the oil from the supply tank from seeping into the crankcase during inoperative periods.
C— To prevent the oil from the pressure pump from entering the scavenger system.

The check valve in the dry-sump lubrication system of a reciprocating engine prevents oil from the tank from seeping into the crankcase during the times when the engine is not operating.

8462. From the following, identify the factor that has the least effect on the oil consumption of a specific engine.

A— Mechanical efficiency.
B— Engine RPM.
C— Lubricant characteristics.

The mechanical efficiency of an engine (the ratio of the power delivered to the propeller shaft to the power produced in the engine) has the least effect on the oil consumption of any of the other items listed in this question.

The engine RPM and the characteristics of the lubricant definitely affect the oil consumption of an aircraft engine.

8463. How is the oil collected by the piston oil ring returned to the crankcase?

A— Down vertical slots cut in the piston wall between the piston oil ring groove and the piston skirt.
B— Through holes drilled in the piston oil ring groove.
C— Through holes drilled in the piston pin recess.

The oil collected by the oil control ring on a piston is returned to the crankcase through holes drilled in the piston ring groove.

8464. Which of the following lubrication system components is never located between the pressure pump and the engine pressure system?

A— Oil temperature bulb.
B— Fuel line for oil dilution system.
C— Check valve.

The fuel line where the gasoline is directed into the lubricating oil to dilute it for cold-weather starting is connected into the lubricating system where the oil pressure is the same as it is inside the oil tank.

Fuel is never fed into the oil between the pressure pump and the pressure system.

8465. As an aid to cold-weather starting, the oil dilution system thins the oil with

A— kerosene.
B— alcohol.
C— gasoline.

Some aircraft engines are equipped with an oil dilution system that puts gasoline from the aircraft fuel tanks into the engine oil before the engine is shut down at night.

The oil is thinned, or diluted, and the friction from the cold oil is greatly reduced for cold-weather starting.

When the engine starts and the oil warms up, the gasoline evaporates from it.

8466. The basic oil pressure relief valve setting for a newly overhauled engine is made

A— within the first 30 seconds of engine operation.
B— when the oil is at a higher than normal temperature to assure high oil pressure at normal oil temperature.
C— in the overhaul shop.

The initial oil-pressure relief valve setting for a newly overhauled engine is made in the overhaul shop.

If the pressure is not correct when the engine is started, it can be adjusted.

8467. Where is the oil temperature bulb located on a dry sump reciprocating engine?

A— Oil inlet line.
B— Oil cooler.
C— Oil outlet line.

The oil-temperature bulb is located in the line between the oil tank and the inlet of the pressure pump on a dry-sump reciprocating engine.

Oil temperature measured at this point is oil-inlet temperature.

Answers

8460 [A] (K03) AMT-P	8461 [B] (K03) AMT-P	8462 [A] (K03) AMT-P	8463 [B] (K03) AMT-P
8464 [B] (K03) AMT-P	8465 [C] (K03) AMT-P	8466 [C] (K03) AMT-P	8467 [A] (K03) AMT-P

8468. Cylinder walls are usually lubricated by

A—splashed or sprayed oil.
B—a direct pressure system fed through the crankshaft, connecting rods, and the piston pins to the oil control ring groove in the piston.
C—oil that is picked up by the oil control ring when the piston is at bottom center.

The cylinder walls of an aircraft reciprocating engine are lubricated by oil that is either sprayed on them from the main bearing of the crankshaft or else splashed on them.

8469. If a full-flow oil filter is used on an aircraft engine, and the filter becomes completely clogged, the

A—oil supply to the engine will be blocked.
B—oil will be bypassed back to the oil tank hopper where larger sediments and foreign matter will settle out prior to passage through the engine.
C—bypass valve will open and the oil pump will supply unfiltered oil to the engine.

In case a full-flow oil filter clogs, provisions are made (usually by a bypass valve) to provide oil to the bearings.

If the filter clogs, the bypass valve will open and allow the pump to supply the engine with unfiltered oil.

It is much better to have unfiltered oil going through the engine than for the engine to have no oil.

8470. Oil accumulation in the cylinders of an inverted inline engine and in the lower cylinders of a radial engine is normally reduced or prevented by

A—reversed oil control rings.
B—routing the valve-operating mechanism lubricating oil to a separate scavenger pump.
C—extended cylinder skirts.

Cylinders of inverted in-line engines and radial engines have skirts that extend up into the crankcase far enough to prevent oil flooding the lower cylinders.

These extended skirts do not prevent oil from getting into the cylinders, but they do minimize it.

Engines with cylinders extending below the center line, should be pulled through by hand before they are started to be sure that there is no oil in any of the lower cylinders.

8471. What is the primary purpose of changing aircraft engine lubricating oils at predetermined periods?

A—The oil becomes diluted with gasoline washing past the pistons into the crankcase.
B—The oil becomes contaminated with moisture, acids, and finely divided suspended solid particles.
C—Exposure to heat and oxygen causes a decreased ability to maintain a film under load.

Oil used in an aircraft engine becomes contaminated with gasoline, moisture, acids, dirt, carbon and metallic particles.

Because of the accumulation of these harmful substances, the oil used in an engine should be periodically drained and replaced with fresh oil.

8472. What determines the minimum particle size which will be excluded or filtered by a cuno-type (stacked disc, edge filtration) filter?

A—The disc thickness.
B—The spacer thickness.
C—Both the number and thickness of the discs in the assembly.

A Cuno filter separates contaminants from the oil by passing the oil between the disks in a stack.

The thickness of the spacers between the disks determines the separation of the disks. This separation determines the minimum particle size that can be filtered out of the oil.

8473. What is the primary purpose of the hopper located in the oil supply tank of some dry sump engine installations?

A—To reduce the time required to warm the oil to operating temperatures.
B—To reduce surface aeration of the hot oil and thus reduce oxidation and the formation of sludge and varnish.
C—To impart a centrifugal motion to the oil entering the tank so that the foreign particles in the oil will separate more readily.

A hopper in the oil supply tank of some dry-sump engine installations reduces the time needed to warm the oil up to its operating temperature. When the oil is cold, the oil in the hopper, rather than in the entire tank, is circulated through the engine.

As the oil warms up, the oil surrounding the hopper gradually feeds into the oil flowing through the engine.

When it is necessary to dilute the oil, gasoline is added so it mixes with the circulating oil. This diluted oil remains in the hopper where it is the first oil used when the engine is started.

Answers

8468 [A] (K03) AMT-P	8469 [C] (K03) AMT-P	8470 [C] (K03) AMT-P	8471 [B] (K03) AMT-P
8472 [B] (K03) AMT-P	8473 [A] (K03) AMT-P		

8474. The purpose of the flow control valve in a reciprocating engine oil system is to

A—direct oil through or around the oil cooler.
B—deliver cold oil to the hopper tank.
C—compensate for volumetric increases due to foaming of the oil.

The flow control valve in an oil cooler used on an aircraft reciprocating engine determines which of two possible paths the oil takes as it passes through the oil cooler.

If the oil is hot, the valve directs it through the core of the cooler where the heat can be removed.

If the oil is not hot, it is bypassed around the outside of the core of the cooler.

8475. Where are sludge chambers, when used in aircraft engine lubrication systems, usually located?

A—In the crankshaft throws.
B—Adjacent to the scavenger pumps.
C—In the oil storage tank.

Sludge chambers are used inside the crankshaft throws of a reciprocating engine.

Sludge is thrown by centrifugal force away from the oil passages in these chambers. It is held there until the engine is overhauled.

8476. Why is an aircraft reciprocating engine oil tank on a dry sump lubrication system equipped with a vent line?

A—To prevent pressure buildup in the reciprocating engine crankcase.
B—To eliminate foaming in the oil tank.
C—To prevent pressure buildup in the oil tank.

Oil tanks used in dry-sump lubrication systems of aircraft reciprocating engines are vented to the engine. The engine crankcase is then vented to the outside air.

This venting system prevents the heating and expansion of the oil from causing a buildup of air pressure inside the oil tank.

8477. Excessive oil is prevented from accumulating on the cylinder walls of a reciprocating engine by

A—the design shape of the piston skirt.
B—internal engine pressure bleeding past the ring grooves.
C—oil control rings on the pistons.

Oil control rings on the pistons of a reciprocating engine control the amount of oil allowed to remain on the cylinder walls.

8478. (1) Wet sump oil systems are most commonly used in gas turbine engines.

(2) In most turbine engine oil tanks, a slight pressurization of the tank is desired to ensure a positive flow of oil.

Regarding the above statements,

A—both No. 1 and No. 2 are true.
B—only No. 2 is true.
C—neither No. 1 nor No. 2 is true.

Statement (1) is not true. Wet-sump lubrication systems are not commonly used on gas turbine engines.

Statement (2) is true. Most turbine engine oil tanks are pressurized to about 4 psi to ensure a positive flow of oil to the pump and prevent cavitation.

8479. The pumping capacity of the scavenger pump in a dry sump aircraft engine's lubrication system

A—is greater than the capacity of the oil supply pump.
B—is less than the capacity of the oil supply pump.
C—is usually equal to the capacity of the oil supply pump in order to maintain constant oiling conditions.

The scavenger pump in a dry-sump aircraft-engine lubricating system has a greater capacity than the pressure pump.

The oil returned to the tank by the scavenger pump has a greater volume than the oil moved by the pressure pump because it is hotter and has air trapped in it.

8480. In which of the following situations will the oil cooler automatic bypass valve be open the greatest amount?

A—Engine oil above normal operating temperature.
B—Engine oil below normal operating temperature.
C—Engine stopped with no oil flowing after runup.

When the engine oil is below its normal operating temperature, the automatic bypass valve is wide open. This wide-open valve allows the oil to completely bypass the cooler.

As the oil warms up, the automatic bypass valve closes and directs the oil through the core of the cooler so it can be cooled.

Answers

8474 [A] (K03) AMT-P
8475 [A] (K03) AMT-P
8476 [C] (K03) AMT-P
8477 [C] (K03) AMT-P
8478 [B] (K03) AMT-P
8479 [A] (K03) AMT-P
8480 [B] (K03) AMT-P

8481. In order to maintain a constant oil pressure as the clearances between the moving parts of an engine increase through normal wear, the supply pump output

A—increases as the resistance offered to the flow of oil increases.
B—remains relatively constant (at a given RPM) with less oil being returned to the pump inlet by the relief valve.
C—remains relatively constant (at a given RPM) with more oil being returned to the pump inlet by the relief valve.

In order to maintain a constant oil pressure as the clearances between the moving parts of an engine increase through normal wear, the output of the oil pressure pump remains relatively constant for any given RPM. But, since there is more oil flowing out through the bearings and other moving parts, less oil is returned to the inlet side of the pump by the relief valve.

8482. The valve assemblies of opposed reciprocating engines are lubricated by means of a

A—gravity feed system.
B—splash and spray system.
C—pressure system.

The valve assemblies of opposed reciprocating engines are lubricated by the pressure system.

Oil under pressure flows through the hydraulic tappet bodies and through hollow pushrods to the rocker arms where it lubricates the rocker arm bearings and the valve stems.

The oil then drains back into the crankcase through either the pushrod housing or an external oil line from the rocker box to the crankcase.

8483. What will result if an oil filter becomes completely blocked?

A—Oil will flow at a reduced rate through the system.
B—Oil flow to the engine will stop.
C—Oil will flow at the normal rate through the system.

Federal Aviation Regulation Part 33 "Airworthiness Standards for Aircraft Engines" requires that an aircraft-engine oil strainer must be constructed and installed in such a way that oil will flow through the engine at its normal rate, even though the strainer or filter element is completely blocked.

8484. A turbine engine dry sump lubrication system of the self-contained, high-pressure design

A—has no heat exchanger.
B—consists of pressure, breather, and scavenge subsystems.
C—stores oil in the engine crankcase.

A turbine-engine dry-sump lubrication system of the self-contained, high-pressure type, such as is used on the Pratt & Whitney JT-3D engine, consists of three basic subsystems: The pressure subsystem supplies oil to the main engine bearings and the accessory drives, the scavenger subsystem scavenges oil from the bearing compartments and the accessory drives, and the breather subsystem interconnects the individual bearing compartments and the lubricating oil tank.

8485. Lube system last chance filters in turbine engines are usually cleaned

A—during annual inspection.
B—during 100-hour inspections.
C—during overhaul.

Last-chance oil filters are installed in the oil lines to prevent plugging of the oil jets.

Last-chance filters are accessible only during engine overhaul.

8486. How are the piston pins of most aircraft engines lubricated?

A—By pressure oil through a drilled passageway in the heavy web portion of the connecting rod.
B—By oil which is sprayed or thrown by the master or connecting rods.
C—By the action of the oil control ring and the series of holes drilled in the ring groove directing oil to the pin and piston pin boss.

Piston pins in an aircraft reciprocating engine are lubricated by oil sprayed or thrown out by the master rod or connecting rod into the inside of the piston.

Oil thrown up into the piston lubricates the piston pin in both the piston and in the small end of the connecting rod.

8487. The vent line connecting the oil supply tank and the engine in some dry sump engine installations permits

A—pressurization of the oil supply to prevent cavitation of the oil supply pump.
B—oil vapors from the engine to be condensed and drained into the oil supply tank.
C—the oil tank to be vented through the normal engine vent.

Oil supply tanks used with dry-sump engines are normally vented to the engine crankcase. The crankcase is, in turn, vented to the outside air.

This method of venting provides adequate ventilation of the tank and prevents the loss of oil through the vents.

Answers

8481 [B] (K03) AMT-P	8482 [C] (K03) AMT-P	8483 [C] (K03) AMT-P	8484 [B] (K03) AMT-P
8485 [C] (K03) AMT-P	8486 [B] (K03) AMT-P	8487 [C] (K03) AMT-P	

8488. An engine lubrication system pressure relief valve is usually located between the

A—oil cooler and the scavenger pump.
B—scavenger pump and the external oil system.
C—pump and the internal oil system.

The pressure relief valve in an engine lubrication system is located between the pump and the internal oil system to maintain the pressure of the oil being moved by the pump at the pressure required for the operation of the engine.

Flow caused by excess pressure is dumped back into the inlet of the pump.

8489. Where is the oil of a dry sump reciprocating engine exposed to the temperature control valve sensing unit?

A—Oil cooler inlet.
B—Engine outlet.
C—Engine inlet.

The oil-temperature control valve sensing unit is located at the inlet to the oil cooler. Hot oil is routed through the core of the cooler, and cold oil is routed around the outside of the core.

8490. Under which of the following conditions is the oil cooler flow control valve open on a reciprocating engine?

A—When the temperature of the oil returning from the engine is too high.
B—When the temperature of the oil returning from the engine is too low.
C—When the scavenger pump output volume exceeds the engine pump input volume.

The oil-cooler flow control valve is normally open when the oil does not need cooling (when the oil temperature is too low). The oil passes through the open valve and around the oil cooler core.

When the oil temperature becomes high enough that it needs to be cooled, this valve closes and the oil flows through the cooler core where it gives up some of its heat to the air.

8491. The purpose of a relief valve installed in the tank venting system of a turbine engine oil tank is to

A—prevent oil pump cavitation by maintaining a constant pressure on the oil pump inlet.
B—maintain internal tank air pressure at the ambient atmospheric level regardless of altitude or rate of change in altitude.
C—maintain a positive internal pressure in the oil tank after shutdown to prevent oil pump cavitation on engine start.

The relief valve in the oil-tank vent system of a turbojet engine prevents the oil pump from cavitating by maintaining a constant positive pressure on the oil pump inlet.

8492. In a reciprocating engine, oil is directed from the pressure relief valve to the inlet side of the

A—scavenger pump.
B—oil temperature regulator.
C—pressure pump.

The oil-pressure relief valve in a reciprocating engine relieves the oil back to the inlet side of the pressure pump.

8493. If the oil in the oil cooler core and annular jacket becomes congealed, what unit prevents damage to the cooler?

A—Oil pressure relief valve.
B—Airflow control valve.
C—Surge protection valve.

If oil becomes congealed in the core of a cooler because of its low temperature, the surge protection valve in the cooler will open and allow the oil to flow around the core of the cooler through an annular jacket.

The operating engine oil flowing through this jacket will warm up the oil in the core so it will liquify and return to the lubrication system.

8494. The primary source of oil contamination in a normally operating reciprocating engine is

A—metallic deposits as a result of engine wear.
B—atmospheric dust and pollution.
C—combustion deposits due to combustion chamber blow-by and oil migration on the cylinder walls.

The primary source of oil contamination, which is present under all operating conditions, is the combustion deposits due to combustion chamber blow-by and oil that has become overheated by its contact with the cylinder walls.

8495. A drop in oil pressure may be caused by

A—the temperature regulator sticking open.
B—the bypass valve sticking open.
C—foreign material under the relief valve.

If foreign material becomes caught under the relief valve, it will hold the valve off its seat and cause the oil pressure to drop.

Answers

8488 [C] (K03) AMT-P	8489 [A] (K03) AMT-P	8490 [B] (K03) AMT-P	8491 [A] (K03) AMT-P
8492 [C] (K03) AMT-P	8493 [C] (K03) AMT-P	8494 [C] (K03) AMT-P	8495 [C] (K03) AMT-P

8496. The main oil filters strain the oil at which point in the system?

A—Immediately after it leaves the scavenger pump.
B—Immediately before it enters the pressure pump.
C—Just as it leaves the pressure pump.

The main oil filter in an aircraft engine strains the oil just after it leaves the pressure pump.

8497. Which type valve prevents oil from entering the main accessory case when the engine is not running?

A—Bypass.
B—Relief.
C—Check.

A check valve in an engine lubricating system prevents oil from the tank from entering the accessory case when the engine is not running.

8498. An oil tank having a capacity of 5 gallons must have an expansion space of

A—2 quarts.
B—4 quarts.
C—5 quarts.

An oil tank must have an expansion space of not less than 10% of the tank capacity, or one-half gallon.

The required expansion space for a five-gallon oil tank is one-half gallon, which is two quarts.

8499. As a general rule, a small amount of small fuzzy particles or gray metallic paste on a turbine engine magnetic chip detector

A—is considered to be the result of normal wear.
B—indicates an imminent component failure.
C—indicates accelerated generalized wear.

Small fuzzy particles or gray metallic paste on the chip detector is considered satisfactory, and is the result of normal wear. But, metallic chips or flakes are an indication of serious internal wear or breakage.

8500. Why is expansion space required in an engine oil supply tank?

A—To eliminate oil foaming.
B—For oil enlargement and collection of foam.
C—For proper oil tank ventilation.

An expansion space is required in an engine oil tank because the oil expands (enlarges) when it gets hot and when it collects foam.

8501. The purpose of a dwell chamber in a turbine engine oil tank is to provide

A—a collection point for sediments.
B—for a pressurized oil supply to the oil pump inlet.
C—separation of entrained air from scavenged oil.

A dwell chamber, or de-areator, is a section of a turbine engine oil tank in which any air that is entrained in the scavenge oil is separated from the oil.

8502. Which of the following bearing types must be continuously lubricated by pressure oil?

A—Ball.
B—Roller.
C—Plain.

Friction-type bearings, which are usually called plain bearings, must be continuously lubricated by pressure oil.

8503. When a magneto is disassembled, keepers are usually placed across the poles of the rotating magnet to reduce the loss of magnetism. These keepers are usually made of

A—chrome magnet steel.
B—soft iron.
C—cobalt steel.

If the magnet is left outside a magneto, it will lose some of its magnetic strength unless its magnetic circuit is completed with a soft iron keeper.

Many of the newer magnets have such a high retentivity that this is no longer the problem it has been in the past.

8504. How is the strength of a magneto magnet checked?

A—Hold the points open and check the output of the primary coil with an ac ammeter while operating the magneto at a specified speed.
B—Check the ac voltage reading at the breaker points.
C—Check the output of the secondary coil with an ac ammeter while operating the magneto at a specified speed.

A magneto is essentially an AC generator whose output is determined by the speed of the rotor and the strength of the permanent magnet.

To test the strength of the permanent magnet, the magneto is mounted on a test stand and driven at a specified speed.

The breaker points are held open and the primary current is measured with an AC ammeter.

Answers

8496 [C] (K03) AMT-P
8497 [C] (K03) AMT-P
8498 [A] (K03) 14 CFR 23.1013
8499 [A] (K03) AMT-P
8500 [B] (K03) AMT-P
8501 [C] (K03) AMT-P
8502 [C] (K03) AMT-P
8503 [B] (L01) AMT-P
8504 [A] (L01) AMT-P

8505. The E-gap angle is usually defined as the number of degrees between the neutral position of the rotating magnet and the position

A—where the contact points close.
B—where the contact points open.
C—of greatest magnetic flux density.

By definition, the E-gap angle is the number of degrees beyond the neutral position of the rotating magnet at which the breaker contact points open.

It is at this point that the flux change is the greatest and the primary current is the greatest.

8506. The greatest density of flux lines in the magnetic circuit of a rotating magnet-type magneto occurs when the magnet is in what position?

A—Full alignment with the field shoe faces.
B—A certain angular displacement beyond the neutral position, referred to as E-gap angle or position.
C—The position where the contact points open.

The greatest flux density in the magnetic circuit of a rotating magnet-type magneto occurs when the magnet is in its full-register position.

In the full-register position, the magnet is fully aligned with its pole shoe faces.

8507. Magneto breaker point opening relative to the position of the rotating magnet and distributor rotor (internal timing) can be set most accurately

A—during the magneto-to-engine timing operation.
B—during assembly of the magneto before installation on the engine.
C—by setting the points roughly at the required clearance before installing the magneto and then making the fine breaker point adjustment after installation to compensate for wear in the magneto drive train.

Internal timing of a magneto is the timing of the opening of the breaker points relative to the position of the rotating magnet and the distributor rotor.

Internal timing is best accomplished while assembling the magneto before it is installed on the engine.

8508. Why are high-tension ignition cables frequently routed from the distributors to the spark plugs in flexible metallic conduits?

A—To eliminate high altitude flashover.
B—To reduce the formation of corona and nitric oxide on the cable insulation.
C—To reduce the effect of the high-frequency electromagnetic waves emanated during operation.

A high-tension ignition cable is "shielded" when it is routed inside a flexible metallic conduit or braid.

The electromagnetic field radiating from the conductor during the time the spark is occurring at the spark plugs, is picked up by this shielding and carried to ground so it cannot cause radio interference.

8509. What will be the results of increasing the gap of the breaker points in a magneto?

A—Retard the spark and increase its intensity.
B—Advance the spark and decrease its intensity.
C—Retard the spark and decrease its intensity.

If the gap of the breaker points is increased, the spark will occur early (it will be advanced). And, because the magneto will be out of time internally, the spark will also be weak.

8510. What is the purpose of a safety gap in some magnetos?

A—To discharge the secondary coil's voltage if an open occurs in the secondary circuit.
B—To ground the magneto when the ignition switch is off.
C—To prevent flashover in the distributor.

Safety gaps are in the secondary circuit of a magneto to protect the insulation of the secondary coil.

If there is an open in the secondary circuit (such as could be caused by a lead being left off a spark plug), the voltage will build up high enough to cause a spark to jump across the safety gap rather than allowing the voltage to rise high enough to damage the secondary coil insulation.

8511. When timing a magneto internally, the alignment of the timing marks indicates that the

A—breaker points are just closing.
B—magnets are in the neutral position.
C—magnets are in the E-gap position.

When the timing marks are aligned while internally timing a magneto, the magnet is in its E-gap position.

The E-gap position is a few degrees of magnet rotation beyond its neutral position, in the direction of normal magnet rotation.

Answers

8505 [B] (L01) AMT-P	8506 [A] (L01) AMT-P	8507 [B] (L01) AMT-P	8508 [C] (L01) AMT-P
8509 [B] (L01) AMT-P	8510 [A] (L01) AMT-P	8511 [C] (L01) AMT-P	

8512. When internally timing a magneto, the breaker points begin to open when the rotating magnet is

A—fully aligned with the pole shoes.
B—a few degrees past full alignment with the pole shoes.
C—a few degrees past the neutral position.

The breaker points in a magneto begin to open when the rotating magnet is in its E-gap position.

The E-gap position is a few degrees of magnet rotation past its neutral position.

8513. What is the electrical location of the primary capacitor in a high-tension magneto?

A—In parallel with the breaker points.
B—In series with the breaker points.
C—In series with the primary and secondary winding.

The primary capacitor in a high-tension magneto is in parallel with the primary coil, the ignition switch and the breaker points.

All three of these components are connected between ground and the ungrounded end of the primary coil.

8514. In a high-tension ignition system, the current in the magneto secondary winding is

A—conducted from the primary winding via the discharge of the capacitor.
B—induced when the primary circuit is interrupted.
C—induced when the primary circuit discharges via the breaker points.

Current is induced into the secondary winding of a magneto coil when the current flowing in the primary winding is interrupted.

When the primary current is interrupted, the magnetic field surrounding the secondary winding collapses and induces a voltage that causes secondary current to flow.

8515. When a "Shower of Sparks" ignition system is activated at an engine start, a spark plug fires

A—as soon as the advance breaker points open.
B—only while both the retard and advance breaker points are closed.
C—only while both the retard and advance breaker points are open.

The left magneto used in a "Shower of Sparks" ignition system has two sets of breaker points connected in parallel. One set is timed to open at the normal time, and the other set opens later, at the correct retarded position for engine starting.

Low-voltage, pulsating DC from the starting vibrator is directed into the primary winding when the ignition/starter switch is placed in the start position. When both sets of points are closed, the pulsating DC flows to ground through the points and nothing happens. When the normal, or run, points open, the current continues to flow to ground through the retard points, and still nothing happens. But, when the retard points open, current can flow to ground only through the primary winding and this induces a high voltage into the secondary winding which produces a continuous spark across the spark plug electrodes as long as both sets of points are open.

8516. What is the radial location of the two north poles of a four-pole rotating magnet in a high-tension magneto?

A—180° apart.
B—270° apart.
C—90° apart.

The two north poles in a four-pole rotating magnet are 180° apart.

Every other pole is opposite, and the flux reverses its direction every 90° of magnet rotation.

8517. Magneto pole shoes are generally made of

A—laminations of high-grade soft iron.
B—laminations of high-grade Alnico.
C—pieces of high-carbon iron.

The pole shoes of a magneto must be made of a material that is highly permeable, yet one that does not retain its magnetism. Most are made of laminations of high-grade soft iron.

Soft iron conducts the lines of magnetic flux with the fewest losses. And, because it is laminated, it reduces the flow of eddy currents that cause electrical losses in the magneto.

8518. Capacitance afterfiring in most modern spark plugs is reduced by the use of

A—fine wire electrodes.
B—a built-in resistor in each plug.
C—aluminum oxide insulation.

Shielded ignition cable acts as a capacitor made of the inner conductor, the insulator, and the metallic shielding. When the spark jumps the gap in the spark plug, energy is stored in this capacitor and as the spark tries to die down it is sustained by this stored energy and capacitance afterfiring results.

Spark plugs used with shielded harnesses have a resistor built into them to quench this sustained spark. The resistor stops the flow of current when the voltage drops to a specified value. This shortens the duration of the spark, and extends the electrode life.

Answers

8512 [C] (L01) AMT-P	8513 [A] (L01) AMT-P	8514 [B] (L01) AMT-P	8515 [C] (L01) AMT-P
8516 [A] (L01) AMT-P	8517 [A] (L01) AMT-P	8518 [B] (L01) AMT-P	

8519. What components make up the magnetic system of a magneto?

A—Pole shoes, the pole shoe extensions, and the primary coil.
B—Primary and secondary coils.
C—Rotating magnet, the pole shoes, the pole shoe extensions, and the coil core.

The magnetic system, or circuit, of a magneto is made up of the rotating magnet, the pole shoes and their extensions, and the coil core.

These components provide a complete circuit for the lines of magnetic flux.

8520. In an aircraft ignition system, one of the functions of the capacitor is to

A—regulate the flow of current between the primary and secondary coil.
B—facilitate a more rapid collapse of the magnetic field in the primary coil.
C—stop the flow of magnetic lines of force when the points open.

When the breaker points in the primary circuit of a magneto begin to open, the resistance of the circuit begins to increase. At this same time, the primary current "sees" what appears to be a low-resistance path to ground through the condenser.

The current flows into the condenser, rather than attempting to flow across the points. This flow aids in the rapid collapse of the magnetic field in the primary coil.

8521. When will the voltage in the secondary winding of a magneto, installed on a normally operating engine, be at its highest value?

A—Just prior to spark plug firing.
B—Toward the latter part of the spark duration when the flame front reaches its maximum velocity.
C—Immediately after the breaker points close.

The magnetic field, caused by the primary current, collapses when the breaker points open. As this magnetic field collapses, it induces a voltage in the secondary winding of the magneto coil.

The voltage in the secondary winding increases until it reaches a value high enough to cause a spark to jump the gap in the spark plug.

The highest voltage in the secondary winding of a magneto coil occurs just before the spark plug fires.

8522. When the switch is off in a battery ignition system, the primary circuit is

A—grounded.
B—opened.
C—shorted.

When the switch in a battery ignition system is turned to the OFF position, the primary circuit is open and no primary current can flow.

8523. As an aircraft engine's speed is increased, the voltage induced in the primary coil of the magneto

A—remains constant.
B—increases.
C—varies with the setting of the voltage regulator.

The amount of voltage induced into the primary coil of a magneto is determined by the rate at which the lines of magnetic flux are cut.

The greater the engine speed, the faster the magneto turns and the faster the lines of flux are cut.

As the engine speed increases, the voltage induced in the primary winding of the magneto increases.

8524. When internally timing a magneto, the breaker points begin to open when

A—the piston has just passed TDC at the end of the compression stroke.
B—the magnet poles are a few degrees beyond the neutral position.
C—the magnet poles are fully aligned with the pole shoes.

When a magneto is properly timed internally, the breaker points should just begin to open at the E-gap position.

This is the position in which there is no magnetic flux in the core of the coil. The rotating magnet is just a few degrees of rotation beyond its neutral position. The direction of flux through the coil core is now in the process of reversing.

8525. The purpose of a safety gap in a magneto is to

A—prevent burning out the primary winding.
B—protect the high-voltage winding from damage.
C—prevent burning of contact points.

Safety gaps are in the secondary circuit for the purpose of protecting the insulation of the secondary coil (the high-voltage winding).

If there is an open in the secondary circuit (such as could be caused by a lead being left off a spark plug), the voltage will build up high enough to cause a spark to jump across the safety gap rather than allowing the voltage to rise high enough to damage the secondary coil insulation.

Answers
8519 [C] (L01) AMT-P
8520 [B] (L01) AMT-P
8521 [A] (L01) AMT-P
8522 [B] (L01) AMT-P
8523 [B] (L01) AMT-P
8524 [B] (L01) AMT-P
8525 [B] (L01) AMT-P

8526. A defective primary capacitor in a magneto is indicated by

A—a fine-grained frosted appearance of the breaker points.
B—burned and pitted breaker points.
C—a weak spark.

A defective primary capacitor in a magneto will allow arcing between the breaker points. This will cause the points to burn and pit.

8527. How many secondary coils are required in a low-tension ignition system on an 18-cylinder engine?

A—36.
B—18.
C—9.

A low-tension ignition system requires a secondary coil for each spark plug.

An 18-cylinder engine will therefore require 36 secondary coils.

8528. A magneto ignition switch is connected

A—in series with the breaker points.
B—parallel to the breaker points.
C—in series with the primary capacitor and parallel to the breaker points.

The magneto ignition switch is connected in parallel with the breaker points so the primary current will flow to ground rather than being interrupted by the breaker points.

8529. The spark is produced in a magneto ignition system when the breaker points are

A—fully open.
B—beginning to open.
C—fully closed.

The sparks are produced in a magneto ignition system as the breaker points just begin to open and interrupt the flow of primary current.

8530. Shielding is used on spark plug and ignition wires to

A—protect the wires from short circuits as a result of chafing or rubbing.
B—prevent outside electromagnetic emissions from disrupting the operation of the ignition system.
C—prevent interference with radio reception.

Spark plug wires and ignition wires are shielded (enclosed in a metal braid) to intercept any electromagnetic energy radiated from these wires. This prevents interference with radio reception.

8531. What is the purpose of using an impulse coupling with a magneto?

A—To absorb impulse vibrations between the magneto and the engine.
B—To compensate for backlash in the magneto and the engine gears.
C—To produce a momentary high rotational speed of the magneto.

An impulse coupling on a magneto spins the magnet at a high rotational speed to produce a hot and retarded spark for starting the engine.

As soon as the engine starts, the coupling disengages, allowing the magneto to fire at its normal advanced position.

8532. The purpose of staggered ignition is to compensate for

A—short ignition harness.
B—rich fuel/air mixture around exhaust valve.
C—diluted fuel/air mixture around exhaust valve.

Staggered ignition timing compensates for the dilution of the fuel-air charge in the cylinder around the exhaust valve.

Igniting the diluted charge before the undiluted charge allows the flame fronts produced by the two spark plugs to meet at the center of the piston head.

8533. Aircraft magneto housings are usually ventilated in order to

A—prevent the entrance of outside air which may contain moisture.
B—allow heated air from the accessory compartment to keep the internal parts of the magneto dry.
C—provide cooling and remove corrosive gases produced by normal arcing.

Magneto housings are normally vented to provide good circulation of air within the magneto.

This ventilation provides some cooling and removes the corrosive gases produced by the normal arcing across the distributor air gap.

8534. Failure of an engine to cease firing after turning the magneto switch off is an indication of

A—an open high tension lead.
B—an open P-lead to ground.
C—a grounded magneto switch.

Failure of an engine to cease firing after turning the magneto switch OFF is an indication of an open in the P-lead between the magneto and ground.

The magneto switch cannot ground the magnetos.

Answers

8526 [B] (L01) AMT-P	8527 [A] (L01) AMT-P	8528 [B] (L01) AMT-P	8529 [B] (L01) AMT-P
8530 [C] (L01) AMT-P	8531 [C] (L01) AMT-P	8532 [C] (L01) AMT-P	8533 [C] (L01) AMT-P
			8534 [B] (L01) AMT-P

8535. Alignment of the marks provided for internal timing of a magneto indicates that the

A—breaker points are just beginning to close for No. 1 cylinder.
B—magneto is in E-gap position.
C—No. 1 cylinder is on TDC of compression stroke.

The timing marks (the marks provided for internal timing) in a magneto are aligned when the rotating magnet is in its E-gap position.

8536. When using a timing light to time a magneto to an aircraft engine, the magneto switch should be placed in the

A—BOTH position.
B—OFF position.
C—LEFT or RIGHT position (either one).

When using a timing light to time a magneto to an aircraft engine, the ignition switch should be placed in the BOTH position.

When the ignition switch is in the BOTH position, the ground circuits of both magnetos are open and the timing light can detect when the breaker points open.

8537. What is the difference between a low-tension and a high-tension engine ignition system?

A—A low-tension system produces relatively low voltage at the spark plug as compared to a high-tension system.
B—A high-tension system is designed for high-altitude aircraft, while a low-tension system is for low- to medium-altitude aircraft.
C—A low-tension system uses a transformer coil near the spark plugs to boost voltage, while the high-tension system voltage is constant from the magneto to the spark plugs.

Low-tension ignition systems are used with some aircraft engines to prevent arcing inside the distributor at high altitudes.

Low-tension ignition systems differ from high-tension systems in that there is no high voltage in the magnetos or in the distributors. Transformer coils mounted on the cylinder heads boost the voltage at each spark plug.

In a high-tension ignition system, the high voltage is produced inside the magneto and is carried through the distributor to the spark plug by way of high-tension ignition leads.

8538. What test instrument could be used to test an ignition harness for suspected leakage?

A—A high tension lead tester.
B—A high voltage dc voltmeter.
C—A high amperage dc ammeter.

A high-tension lead tester is used to test an ignition harness for suspected electrical leakage.

A high DC voltage is placed across the insulation, and some form of current indication shows whether or not there is leakage.

8539. The amount of voltage generated in any magneto secondary coil is determined by the number of windings and by the

A—rate of buildup of the magnetic field around the primary coil.
B—rate of collapse of the magnetic field around the primary coil.
C—amount of charge released by the capacitor.

The amount of voltage induced in the secondary winding of a magneto coil is determined by two things: by the number of turns of wire in the secondary winding and by the rate of collapse of the magnetic field around the primary coil.

8540. Magneto breaker points must be timed to open when the

A—rotating magnet is positioned a few degrees before neutral.
B—greatest magnetic field stress exists in the magnetic circuit.
C—rotating magnet is in the full register position.

Magneto breaker points are timed to open when the magnetic field stress in the magnetic circuit is greatest.

The magnetic field stress is caused by the difference in the resultant magnetic flux produced by the rotating magnet and the primary current, and the static flux produced by the rotating magnet alone.

Opening the points at the instant of the greatest magnetic field stress produces the highest voltage in the secondary winding.

Answers

8535 [B] (L01) AMT-P	8536 [A] (L01) AMT-P	8537 [C] (L01) AMT-P	8538 [A] (L01) AMT-P
8539 [B] (L01) AMT-P	8540 [B] (L01) AMT-P		

8541. In reference to a "Shower of Sparks" ignition system,

(1) the retard breaker points are designed to keep the affected ignition system operating if the advance breaker points should fail during normal engine operation (after start).

(2) the timed opening of the retard breaker points is designed to prevent engine "kickback" during start.

Regarding the above statements,

A—only No. 1 is true.
B—only No. 2 is true.
C—both No. 1 and No. 2 are true.

Statement (1) is not true. The retard breaker points are used to provide a late spark for starting the engine. They are out of the circuit during normal engine operation, after start.

Statement (2) is true. The retard breaker points are timed to open when the pistons have passed top center and started down. This prevents the engine kicking back when it is being started.

8542. The capacitor-type ignition system is used almost universally on turbine engines primarily because of its high voltage and

A—low amperage.
B—long life.
C—high-heat intensity.

The capacitor-type high-energy ignition system is used on turbine engines because of the high voltage and extremely high heat intensity of the spark it produces.

8543. In a low-tension ignition system, each spark plug requires an individual

A—capacitor.
B—breaker assembly.
C—secondary coil.

A low-tension ignition system differs from a high-tension system in that each spark plug in a low-tension system has its own individual transformer.

Each transformer has a primary and a secondary coil.

8544. A certain nine-cylinder radial engine used a non-compensated single-unit, dual-type magneto with a four-pole rotating magnet and separately mounted distributors. Which of the following will have the lowest RPM at any given engine speed?

A—Breaker cam.
B—Engine crankshaft.
C—Distributors.

All nine cylinders of the engine must fire in 720° of crankshaft rotation.

1. *The four-lobe breaker cam mounts on the four-pole rotating magnet. Since its four lobes must fire nine cylinders, it turns at 1-1/8 times of the crankshaft speed.*
2. *The speed of the crankshaft is the reference from which we are measuring the other speeds.*
3. *The distributors select the spark plug (cylinder) that is to be fired. Since all cylinders are fired in two revolutions of the crankshaft, the distributors turn at one-half the crankshaft speed (they are the slowest turning components).*

8545. What will be the effect if the spark plugs are gapped too wide?

A—Insulation failure.
B—Hard starting.
C—Lead damage.

If spark plugs are gapped too wide, or if the gap has increased due to electrode wear, the engine will likely be hard to start.

The wide gap causes the voltage to have to rise to a higher value before it can produce a spark. In a magneto ignition system, this requires that the magneto turn faster to produce the spark.

8546. When removing a shielded spark plug, which of the following is most likely to be damaged?

A—Center electrode.
B—Shell section.
C—Core insulator.

If a side load is applied to the deep socket wrench used for removing the spark plug, or to the shielded lead as it is taken out of the spark plug, there is a possibility that the core insulator may be damaged.

8547. What likely effect would a cracked distributor rotor have on a magneto?

A—Ground the secondary circuit through the crack.
B—Fire two cylinders simultaneously.
C—Ground the primary circuit through the crack.

If the distributor rotor in a magneto is cracked, carbon and other contaminants will collect in the crack and form a conductive path that will ground the secondary circuit.

The high voltage will find a path to ground through the crack and it will not build up high enough to jump the gap in the spark plug.

Answers

8541 [B] (L02) AMT-P	8542 [C] (L02) AMT-P	8543 [C] (L02) AMT-P	8544 [C] (L02) AMT-P
8545 [B] (L02) AMT-P	8546 [C] (L02) AMT-P	8547 [A] (L02) AMT-P	

8548. How does the ignition system of a gas turbine engine differ from that of a reciprocating engine?

A—One igniter plug is used in each combustion chamber.
B—Magneto-to-engine timing is not critical.
C—A high-energy spark is required for ignition.

Turbine engine ignition systems are required to operate for starting only, but they must provide a high energy discharge at the igniter. An extremely hot spark is needed to relight the engine if it flames out at altitude.

8549. In a turbine engine dc capacitor discharge ignition system, where are the high-voltage pulses formed?

A—At the breaker.
B—At the triggering transformer.
C—At the rectifier.

In the DC capacitor-discharge ignition system, such as is used on some turbojet engines, the storage capacitor is charged to a given voltage.

When this voltage is reached, the gas in the discharge tube ionizes causing a portion of the accumulated charge to flow through the primary of the triggering transformer. This induces a very high voltage in the secondary winding connected to the igniter.

This voltage is high enough to jump the gap and produce the trigger spark, which ionizes the gas in the gap. The storage capacitor discharges the remainder of its accumulated energy through the ionized gap together with the charge from the trigger capacitor.

8550. Which of the following breaker point characteristics is associated with a faulty capacitor?

A—Crowned.
B—Fine grained.
C—Coarse grained.

Properly operating points have a fine-grained frosty or silvery appearance, which should not be confused with the coarse-grained and sooty points caused by faulty condenser action.

8551. How are most radial engine spark plug wires connected to the distributor block?

A—By use of cable-piercing screws.
B—By use of self-locking cable ferrules.
C—By use of terminal sleeves and retaining nuts.

Radial-engine distributors hold the high-voltage leads in their recesses with cable-piercing screws.

This is not generally the case with any of the magnetos used on the modern, horizontally opposed engines.

8552. Thermocouples are usually inserted or installed on the

A—front cylinder of the engine.
B—rear cylinder of the engine.
C—hottest cylinder of the engine.

Thermocouples for measuring cylinder-head temperature are usually attached to a gasket under the spark plug, or are inserted in an adapter in the fins of the head of the hottest-running cylinder in the engine.

8553. Capacitance afterfiring of a spark plug is caused by

A—the stored energy in the ignition shielded lead unloading after normal timed ignition.
B—excessive center electrode erosion.
C—constant polarity firing.

Shielded ignition cable acts as a capacitor made of the inner conductor, the insulator, and the metallic shielding. When the spark jumps the gap in the spark plug, energy is stored in this capacitor and as the spark tries to die down it is sustained by this stored energy and capacitance afterfiring results.

Spark plugs used with shielded harnesses have a resistor built into them to quench this sustained spark. The resistor stops the flow of current when the voltage drops to a specified value. This shortens the duration of the spark, and extends the electrode life.

8554. If it is found that a shielded ignition system does not adequately reduce ignition noise, it may be necessary to install

A—a second layer of shielding.
B—a filter between the magneto and magneto switch.
C—bonding wires from the shielding to ground.

If the shielding does not adequately reduce ignition noise, it may be necessary to install a filter between the magneto and the ignition switch.

The filter blocks radio-frequency noise pulses with a coil and shunts them to ground through a capacitor.

8555. When a magneto is operating, what is the probable cause for a shift in internal timing?

A—The rotating magnet looses its magnetism.
B—The distributor gear teeth are wearing on the rotor gear teeth.
C—The cam follower wear and/or breaker points wear.

Two things may cause the internal timing of a magneto to shift. Wear of the breaker points causes the timing to drift early, and wear of the cam follower causes the timing to drift late. Ideally, these two wear conditions should cancel each other.

Answers

8548 [C] (L02) AMT-P	8549 [B] (L02) AMT-P	8550 [C] (L02) AMT-P	8551 [A] (L02) AMT-P
8552 [C] (L02) AMT-P	8553 [A] (L02) AMT-P	8554 [B] (L02) AMT-G	8555 [C] (L02) AMT-P

8556. Why are turbine engine igniters less susceptible to fouling than reciprocating engine spark plugs?

A—The high-intensity spark cleans the igniter.
B—The frequency of the spark is less for igniters.
C—Turbine igniters operate at cooler temperatures.

Turbine-engine igniters are less susceptible to fouling than reciprocating-engine spark plugs because the high-energy spark jumping the gap cleans the contaminants out of the gap with heat.

8557. The constrained-gap igniter plug used in some gas turbine engines operates at a cooler temperature because

A—it projects into the combustion chamber.
B—the applied voltage is less.
C—the construction is such that the spark occurs beyond the face of the combustion chamber liner.

The constrained-gap igniter plug used in some gas turbine engines operates at a cooler temperature than other types of igniters because it does not project into the combustion chamber liner.

The spark does not remain close to the plug, but arcs beyond the face of the combustion chamber liner.

8558. What should be used to clean grease or carbon tracks from capacitors or coils that are used in magnetos?

A—Solvent.
B—Acetone.
C—Naphtha.

Grease or carbon tracks can be removed from condensers or coils by wiping them with a lint-free cloth moistened with acetone.

8559. Generally, when removing a turbine engine igniter plug, in order to eliminate the possibility of the technician receiving a lethal shock, the ignition switch is turned off and

A—disconnected from the power supply circuit.
B—the igniter lead is disconnected from the plug and the center electrode grounded to the engine after disconnecting the transformer-exciter input lead and waiting the prescribed time.
C—the transformer-exciter input lead is disconnected and the center electrode grounded to the engine after disconnecting the igniter lead from the plug and waiting the prescribed time.

When removing an igniter plug, disconnect the transformer input lead, wait the length of time recommended by the manufacturer, normally one to five minutes, then disconnect the igniter lead and ground the center electrode to the engine. This discharges the capacitor and makes it safe to remove the igniter plug.

8560. Great caution should be exercised in handling damaged hermetically sealed turbine engine igniter transformer units because

A—compounds in the unit may become a fire or explosion hazard when exposed to the air.
B—some contain radioactive material.
C—some contain toxic chemicals.

Some transformer units contain radioactive cesium, and should be handled with extreme care if the hermetic sealing has been damaged.

8561. Igniter plugs used in turbine engines are subjected to high intensity spark discharges and yet they have a long service life because they

A—operate at much lower temperatures.
B—are not placed directly into the combustion chamber.
C—do not require continuous operation.

The service life of an igniter is much longer than the life of a spark plug used in a reciprocating engine because the igniter does not operate continuously.

8562. The electrical circuit from the spark plug back to the magneto is completed by grounding through the

A—engine structure.
B—P-lead.
C—cockpit switch.

After the secondary current flows across the gap in the spark plug, it completes its circuit back to the magneto through the engine structure.

8563. Spark plugs are considered worn out when the

A—electrodes have worn away to about one-half of their original dimensions.
B—center electrode edges have become rounded.
C—electrodes have worn away to about two-thirds of their original dimensions.

Spark plugs are considered to be worn out when their electrodes have worn away to approximately one-half their original dimensions.

Answers

8556 [A] (L02) AMT-P	8557 [C] (L02) AMT-P	8558 [B] (L02) AMT-P	8559 [B] (L02) AMT-P
8560 [B] (L02) AMT-P	8561 [C] (L02) AMT-P	8562 [A] (L02) AMT-P	8563 [A] (L02) AMT-P

8564. Which of the following could cause damage to the nose ceramic or to the electrode of an aircraft spark plug?

A—Plug installed without a copper gasket.
B—Improper gapping procedure.
C—Excessive magneto voltage.

Improper gapping procedure can fracture the nose center ceramic insulator or damage the electrodes of a spark plug.

When gapping a spark plug, use the correct gapping tool and move only the ground electrode.

Do not exert any force against the center electrode.

8565. Sharp bends should be avoided in ignition leads primarily because

A—weak points may develop in the insulation through which high tension current can leak.
B—ignition lead wire conductor material is brittle and may break.
C—ignition lead shielding effectiveness will be reduced.

Sharp bends in ignition leads should be avoided. If a cable is bent sharply or twisted, the insulation is under stress and can develop weak points through which high-tension current can leak.

8566. In a high-tension ignition system, a primary capacitor of too low a capacity will cause

A—excessive primary voltage.
B—excessively high secondary voltage.
C—the breaker contacts to burn.

If the capacity of the primary condenser is too low, it will not be able to absorb all of the surge when the contacts open. There will be some arcing across the breaker contacts and the contacts will burn.

8567. Which of the following, obtained during magneto check at 1,700 RPM, indicates a short (grounded) circuit between the right magneto primary and the ignition switch?

A—BOTH—1,700 RPM,
R—1,625 RPM,
L—1,700 RPM,
OFF—1,625 RPM.
B—BOTH—1,700 RPM,
R—0 RPM,
L—1,700 RPM,
OFF—0 RPM.
C—BOTH—1,700 RPM,
R—0 RPM,
L—1,675 RPM,
OFF—0 RPM.

In the situation here, the right magneto is shorted to ground (grounded) and is therefore not producing any sparks.

When the ignition switch is placed in the R position, the left magneto is grounded. With both magnetos grounded, the engine produces zero RPM.

When the switch is placed in the L position, the right magneto is grounded, but since it was already grounded, the engine still produces 1,700 RPM.

In the OFF position, the left magneto is grounded and the right magneto is already grounded, so the RPM is zero.

In the BOTH position, the left magneto is supplying all of the sparks and the engine is running at 1,700 RPM.

8568. If an aircraft ignition switch is turned off and the engine continues to run normally, the trouble is probably caused by

A—an open ground lead in the magneto.
B—arcing magneto breaker points.
C—primary lead grounding.

If an engine continues to run normally after the ignition switch is turned off, the trouble is an open ground lead in one or both of the magneto circuits.

It does not necessarily have to be in the magneto as the question indicates, but it is in the primary lead (P-lead) circuit.

8569. Which statement is correct regarding the ignition system of a turbine engine?

A—The system is normally de-energized as soon as the engine starts.
B—It is energized during the starting and warmup periods only.
C—The system generally includes a polar inductor-type magneto.

The ignition system of a turbojet engine is operated only during the starting cycle. It is normally de-energized when the engine is running.

Answers

8564 [B] (L02) AMT-P	8565 [A] (L02) AMT-P	8566 [C] (L02) AMT-P	8567 [B] (L02) AMT-P
8568 [A] (L02) AMT-P	8569 [A] (L02) AMT-P		

8570. When the ignition switch of a single (reciprocating) engine aircraft is turned to the OFF position,

A—the primary circuits of both magnetos are grounded.
B—the secondary circuits of both magnetos are opened.
C—all circuits are automatically opened.

When the magneto ignition switch of a single-engine aircraft is open, ON, the primary current produced in the magneto is interrupted by the breaker points and a high voltage is induced into the secondary winding.

When the switch is closed, OFF, the primary coil is grounded. Since the current induced into the primary winding goes directly to ground it is not interrupted and no high voltage is induced into the secondary winding.

When the ignition switch is turned to the OFF position, the primary circuits of both magnetos are grounded.

8571. A spark plug's heat range is the result of

A—the area of the plug exposed to the cooling airstream.
B—its ability to transfer heat from the firing end of the spark plug to the cylinder head.
C—the heat intensity of the spark.

The heat range of a spark plug is a measure of its ability to transfer heat from the center electrode (the firing end) of the spark plug into the cylinder head.

A hot spark plug has a long path for the heat to follow, and it retains heat. A cold spark plug has a short path for the heat, and it transfers its heat.

8572. If staggered ignition timing is used, the

A—spark plug nearest the exhaust valve will fire first.
B—spark will be automatically advanced as engine speed increases.
C—spark plug nearest the intake valve will fire first.

When staggered ignition timing is used in a reciprocating engine, the two sparks in a cylinder occur at different times.

The spark plug on the exhaust side of the cylinder always fires first. The slow rate of burning of the expanded and diluted fuel-air mixture at this point in the cylinder needs an advance in the ignition timing so the flames caused by the two sparks will meet near the center of the piston head.

8573. The term "reach," as applied to spark plug design and/or type, indicates the

A—linear distance from the shell gasket seat to the end of the threads on the shell skirt.
B—length of center electrode exposed to the flame of combustion.
C—length of the shielded barrel.

The reach of a spark plug is the length of the threaded portion of the spark plug that screws into the cylinder head.

This is the linear distance from the shell gasket seat to the end of the shell skirt.

8574. The numbers appearing on the ignition distributor block indicate the

A—sparking order of the distributor.
B—relation between distributor terminal numbers and cylinder numbers.
C—firing order of the engine.

The numbers on a magneto distributor block indicate the sparking order of the distributor.

These numbers must be matched to the firing order of the engine, not to the cylinder numbers.

8575. When testing a magneto distributor block for electrical leakage, which of the following pieces of test equipment should be used?

A—A high-tension harness tester.
B—A continuity tester.
C—A high-range ammeter.

Electrical leakage in a high-voltage component, such as a distributor block, must be checked by using a high-voltage tester, such as a high-tension harness tester.

8576. (1) The platinum and iridium ground electrodes used on fine wire spark plugs are extremely brittle and can be broken if they are improperly handled or adjusted.

(2) When gapping massive-electrode spark plugs, a wire gauge should be inserted between the center and ground electrodes while moving the ground electrode in order to avoid setting the gap too close.

Regarding the above statements,

A—only No. 1 is true.
B—only No. 2 is true.
C—both No. 1 and No. 2 are true.

Statement (1) is true. Fine-wire spark plugs lend themselves to easy gap adjustment. However, caution must be exercised when closing the gap, as the ground electrodes, especially iridium electrodes, are quite brittle and can fracture if they are improperly handled.

Statement (2) is not true. The wire gauge should not *be placed between the electrodes when moving the ground electrode on a massive electrode spark plug. If the wire gauge is between the electrodes, pressure will be applied to the center electrode, and the nose core insulator will be strained and probably cracked.*

Answers

8570 [A] (L02) AMT-P	8571 [B] (L02) AMT-P	8572 [A] (L02) AMT-P	8573 [A] (L02) AMT-P
8574 [A] (L02) AMT-P	8575 [A] (L02) AMT-P	8576 [A] (L02) AMT-P	

8577. Hot spark plugs are generally used in aircraft powerplants

A—with comparatively high compression or high operating temperatures.
B—with comparatively low operating temperatures.
C—which produce high power per cubic inch displacement.

Hot spark plugs (spark plugs with a long path for the heat to follow when traveling from the nose-core insulator into the cylinder head) are used in engines with comparatively low operating temperatures.

The combustion deposits burn off of a hot spark plug rather than fouling the spark plug.

8578. If a spark plug lead becomes grounded, the

A—magneto will not be affected.
B—distributor rotor finger will discharge to the next closest electrode within the distributor.
C—capacitor will break down.

If a spark plug lead is grounded, the high voltage will be bypassed to ground and nothing will happen. The magneto will not be affected but the spark plug will not fire.

8579. Which of the following statements regarding magneto switch circuits is NOT true?

A—In the BOTH position, the right and left magneto circuits are grounded.
B—In the OFF position, neither the right nor left magneto circuits are open.
C—In the RIGHT position, the right magneto circuit is open and the left magneto circuit is grounded.

In this question, we are looking for the alternative that is not *true.*

When a magneto switch is in the BOTH position, the ground circuits for both magnetos are open and neither magneto is grounded.

8580. Which of the following statements most accurately describes spark plug heat range?

A—The length of the threaded portion of the shell usually denotes the spark plug heat range.
B—A hot plug is designed so that the insulator tip is reasonably short to hasten the rate of heat transfer from the tip through the spark plug shell to the cylinder head.
C—A cold plug is designed so that the insulator tip is reasonably short to hasten the rate of heat transfer from the tip through the spark plug shell to the cylinder head.

The heat range of a spark plug relates to the ability of the spark plug to transfer heat from its nose core insulator into the cylinder head.

A cold spark plug has a short insulator tip so that heat can readily transfer from the spark plug nose into the shell and keep the operating temperature of the nose relatively low.

Cold spark plugs are used in high-compression engines where cylinder temperatures are high.

8581. When does battery current flow through the primary circuit of a battery ignition coil?

A—Only when the breaker points are open.
B—At all times when the ignition switch is on.
C—When the breaker points are closed and the ignition switch is on.

Battery current flows through the primary circuit of a battery ignition coil when the ignition switch is ON and the breaker points are closed.

8582. In order to turn a magneto off, the primary circuit must be

A—grounded.
B—opened.
C—shorted.

Current flows in the primary circuit of a magneto any time the magnet rotates. But if the primary circuit is grounded by the ignition switch, this flow is never interrupted by the breaker points.

If the primary current is not interrupted, no high voltage can be induced into the secondary winding.

The magneto is OFF when the primary circuit is grounded.

8583. When performing a magneto ground check on an engine, correct operation is indicated by

A—a slight increase in RPM.
B—no drop in RPM.
C—a slight drop in RPM.

Correct operation of a magneto ignition system is indicated by a slight drop in RPM when the ignition switch is switched from the BOTH position to the position for either magneto by itself. The RPM drop is caused by the inefficiency of the engine operating on one magneto alone.

Answers

8577 [B] (L02) AMT-P	8578 [A] (L02) AMT-P	8579 [A] (L02) AMT-P	8580 [C] (L02) AMT-P
8581 [C] (L02) AMT-P	8582 [A] (L02) AMT-P	8583 [C] (L02) AMT-P	

8584. Defective spark plugs will cause the engine to run rough at

A—high speeds only.
B—low speeds only.
C—all speeds.

Defective spark plugs will cause intermittent missing of the engine at all speeds.

8585. A spark plug is fouled when

A—its spark grounds by jumping electrodes.
B—it causes preignition.
C—its spark grounds without jumping electrodes.

A spark plug is fouled when the high-voltage current passes to ground without causing a spark to jump between the electrodes.

8586. Which of the following would be cause for rejection of a spark plug?

A—Carbon fouling of the electrode and insulator.
B—Insulator tip cracked.
C—Lead fouling of the electrode and insulator.

A spark plug must be rejected if the insulator tip is cracked.

Neither of the other alternatives for this question are cause for spark plug rejection. Spark plugs with these conditions can be cleaned.

8587. What will be the result of using too hot a spark plug?

A—Fouling of plug.
B—Preignition.
C—Burned capacitor.

The use of a spark plug with a heat range that is too hot for the engine will cause a local hot spot within the cylinder.

This can cause preignition.

8588. Upon inspection of the spark plugs in an aircraft engine, the plugs were found caked with a heavy black soot. This indicates

A—worn oil seal rings.
B—a rich mixture.
C—a lean mixture.

A heavy black soot deposit, caked in a spark plug, is an indication that the engine, or at least the cylinder from which the spark plug was removed, has been operating with an excessively rich mixture.

The soot comes from an excess of carbon caused by burning the fuel with too little air.

8589. Spark plug heat range is determined by

A—the reach of the spark plug.
B—its ability to transfer heat to the cylinder head.
C—the number of ground electrodes.

The heat range of a spark plug is determined by its ability to transfer heat into the cylinder head.

A cold spark plug transfers heat most readily, while a hot spark plug has a long nose-core insulator and does not transfer heat as readily.

8590. Ignition check during engine runup indicates excessive RPM drop during operation on the right magneto. The major portion of the RPM loss occurs rapidly after switching to the right magneto position (fast drop). The most likely cause is

A—faulty or fouled spark plugs.
B—incorrect ignition timing on both magnetos.
C—one or more dead cylinders.

An RPM drop that occurs rapidly after switching from both magnetos to the right magneto indicates faulty or fouled spark plugs rather than incorrect ignition or valve timing.

8591. If new breaker points are installed in a magneto on an engine, it will be necessary to time the

A—magneto internally and the magneto to the engine.
B—breaker points to the No. 1 cylinder.
C—magneto drive to the engine.

Any time new breaker points are installed in a magneto, the magneto must be internally timed and timed to the engine.

8592. Using a cold spark plug in a high-compression aircraft engine would probably result in

A—normal operation.
B—a fouled plug.
C—detonation.

A high-compression engine should use a cold spark plug for normal operation.

Cold spark plugs transfer most of the heat into the cylinder head. This prevents a local hot spot inside the cylinder.

Answers

8584 [C] (L02) AMT-P	8585 [C] (L02) AMT-P	8586 [B] (L02) AMT-P	8587 [B] (L02) AMT-P
8588 [B] (L02) AMT-P	8589 [B] (L02) AMT-P	8590 [A] (L02) AMT-P	8591 [A] (L02) AMT-P
			8592 [A] (L02) AMT-P

8593. Spark plug fouling caused by lead deposits occurs most often

A—during cruise with rich mixture.
B—when cylinder head temperatures are relatively low.
C—when cylinder head temperatures are high.

Lead fouling of a spark plug occurs when the cylinder-head temperatures are relatively low and there is not enough heat to vaporize the lead deposits. Unless these deposits are vaporized, they cannot be properly scavenged.

8594. In a four-stroke cycle aircraft engine, when does the ignition event take place?

A—Before the piston reaches TDC on compression stroke.
B—After the piston reaches TDC on power stroke.
C—After the piston reaches TDC on compression stroke.

Ignition occurs in a four-stroke-cycle aircraft engine somewhere in the neighborhood of 20° of crankshaft rotation before the piston reaches top center at the end of the compression stroke.

8595. When installing a magneto on an engine, the

A—piston in the No. 1 cylinder must be a prescribed number of degrees before top center on the compression stroke.
B—magneto breaker points must be just closing.
C—piston in the No. 1 cylinder must be a prescribed number of degrees after top center on the intake stroke.

When installing a magneto on an aircraft engine, the piston in cylinder number 1 must be placed so it is the correct number of degrees of crankshaft rotation before top center on its compression stroke.

8596. The spark occurs at the spark plug when the ignition's

A—secondary circuit is completed.
B—primary circuit is completed.
C—primary circuit is broken.

The spark occurs at the spark plug when the current flowing in the primary circuit of the magneto is interrupted, or broken, by the breaker points opening.

8597. The type of ignition system used on most turbine aircraft engines is

A—high resistance.
B—low tension.
C—capacitor discharge.

The ignition system used on most jet aircraft engines is the high-energy capacitor-discharge type of system.

8598. Ignition check during engine runup indicates a slow drop in RPM. This is usually caused by

A—defective spark plugs.
B—a defective high-tension lead.
C—incorrect ignition timing or valve adjustment.

A slow drop in RPM during an ignition system test, normally indicates incorrect ignition timing or valve adjustment rather than defective spark plugs or high-tension leads.

8599. If the ground wire of a magneto is disconnected at the ignition switch, the result will be the

A—affected magneto will be isolated and the engine will run on the opposite magneto.
B—engine will stop running.
C—engine will not stop running when the ignition switch is turned off.

If the ground wire of a magneto is disconnected at the ignition switch, the engine will not stop running when the ignition switch is turned to the OFF position.

When the ground wire is loose or disconnected, the magneto is said to be "hot."

8600. Which of the following are advantages of dual ignition in aircraft engines?

1. Gives a more complete and quick combustion of the fuel.
2. Provides a backup magneto system.
3. Increases the output power of the engine.
4. Permits the use of lower grade fuels.
5. Increases the intensity of the spark at the spark plugs.

A—2, 3, 4.
B—2, 3, 5.
C—1, 2, 3.

Statement 1 is an advantage of dual ignition. Dual ignition does give a more complete and quick combustion of the fuel.

Statement 2 is an advantage of dual ignition. Dual ignition provides a backup magneto system.

Statement 3 is an advantage of dual ignition. Dual ignition increases the output power of the engine.

Statement 4 is not an advantage of dual ignition.

Statement 5 is not an advantage of dual ignition.

Answers

8593 [B] (L02) AMT-P
8597 [C] (L02) AMT-P
8594 [A] (L02) AMT-P
8598 [C] (L02) AMT-P
8595 [A] (L02) AMT-P
8599 [C] (L02) AMT-P
8596 [C] (L02) AMT-P
8600 [C] (L02) AMT-P

8601. How does high-tension ignition shielding tend to reduce radio interference?

A—Prevents ignition flashover at high altitudes.
B—Reduces voltage drop in the transmission of high-tension current.
C—Receives and grounds high-frequency waves coming from the magneto and high-tension ignition leads.

High-tension ignition cable is enclosed in a metallic braid (shielding) which is grounded by connecting it to the engine.

Radio-frequency electrical energy that radiates from the wire each time a spark jumps the gap in the spark plug is intercepted by the shielding and carried to ground.

Directing this energy to ground prevents it from interfering with communications and navigation radio equipment installed in the aircraft.

8602. Which of the following are distinct circuits of a high-tension magneto?

1. Magnetic.
2. Primary.
3. E-gap.
4. P-lead.
5. Secondary.

A—1, 2, 5.
B—1, 3, 4.
C—2, 4, 5.

A high-tension magneto has a magnetic circuit, a primary circuit and a secondary circuit.

Magnetic flux travels in the magnetic circuit, low-voltage alternating current travels in the primary circuit and pulses of high-voltage alternating current travel in the secondary circuit.

8603. What are two parts of a distributor in an aircraft engine ignition system?

1. Coil.
2. Block.
3. Stator.
4. Rotor.
5. Transformer.

A—2 and 4.
B—3 and 4.
C—2 and 5.

Two major components in the distributor of an aircraft-engine ignition system are the block and the rotor.

The block serves as a terminal to connect all of the spark plug wires to the distributor. The rotor selects the particular wire to which the high voltage from the coil is directed.

8604. What is a result of "flashover" in a distributor?

A—Intense voltage at the spark plug.
B—Reversal of current flow.
C—Conductive carbon trail.

Flashover in the distributor of an aircraft-engine ignition system is a condition in which high voltage leaks from the terminal to ground across the surface of the insulation in the distributor block.

As the high voltage crosses the insulator, it burns, or turns to carbon, any oil or impurity on the surface. This carbon produces a low-resistance path, or conductive trail, for other pulses of voltage to leak across.

8605. What is the relationship between distributor and crankshaft speed of aircraft reciprocating engines?

A—The distributor turns at one-half crankshaft speed.
B—The distributor turns at one and one-half crankshaft speed.
C—The crankshaft turns at one-half distributor speed.

In order for the distributor to direct high voltage to each of the cylinders in two revolutions of the crankshaft, the distributor rotor must turn at one half of the crankshaft speed.

8606. Why do turbine engine ignition systems require high energy?

A—To ignite the fuel under conditions of high altitude and high temperatures.
B—Because the applied voltage is much greater.
C—To ignite the fuel under conditions of high altitude and low temperatures.

Turbine-engine ignition systems require much higher energy than the ignition system for a reciprocating engine. The reason is that turbine igniters must be able to consistently ignite the fuel-air mixture under conditions of high altitude and extremely low temperature.

Answers

8601 [C] (L02) AMT-P	8602 [A] (L02) AMT-P	8603 [A] (L02) AMT-P	8604 [C] (L02) AMT-P
8605 [A] (L02) AMT-P	8606 [C] (L02) AMT-P		

8607. Which of the following are included in a typical turbine engine ignition system?

1. Two igniter plugs.
2. Two transformers.
3. One exciter unit.
4. Two intermediate ignition leads.
5. Two low tension igniter leads.
6. Two high tension igniter leads.

A—2, 3, 4
B—1, 4, 5.
C—1, 3, 6.

A typical turbine-engine ignition system supplies high voltage to two igniters in the combustors on the opposite sides of the engine. In order to do this, the following components are needed:

1. Two igniter plugs
3. One exciter unit (a dual unit)
6. Two high-tension igniter leads

8608. At what RPM is a reciprocating engine ignition switch check made?

A—1,500 RPM.
B—The slowest possible RPM.
C—Full throttle RPM.

The ignition switch check for a reciprocating engine is made at the slowest possible RPM.

The engine is idled and the ignition switch is turned to the OFF position to see if the engine will stop running.

As soon as it is determined that the switch does ground out both magnetos, the switch is returned to the BOTH position.

If the engine is operating at a high RPM when this test is made, there is a possibility of a backfire that will damage the exhaust system.

8609. What is the approximate position of the rotating magnet in a high-tension magneto when the points first close?

A—Full register.
B—Neutral.
C—A few degrees after neutral.

The breaker points in a high-tension magneto open when the rotating magnet is in its E-gap position. In this position, the magnet is just beyond its neutral position.

The breaker points close when the magnet is approximately in its full register position. In this position, the poles of the magnet are aligned with the pole shoes.

8610. What component of a dual magneto is shared by both ignition systems?

A—High-tension coil.
B—Rotating magnet.
C—Capacitor.

A dual, or double magneto, has a single rotating magnet, but it has two sets of breaker points, two high-tension coils and two capacitors.

8611. What would be the result if a magneto breaker point mainspring did not have sufficient tension?

A—The points will stick.
B—The points will not open to the specified gap.
C—The points will float or bounce.

If the breaker-point mainspring (the spring on which the movable breaker point is mounted) does not have sufficient tension, the points will float or bounce. This can cause the engine to miss at high speed.

8612. The secondary coil of a magneto is grounded through the

A—ignition switch.
B—primary coil.
C—grounded side of the breaker points.

The coil of a high-tension magneto is made of a primary coil wound of several layers of fairly heavy wire over a laminated steel core.

One end of the primary coil is grounded to the core. The other end is carried out to the pigtail that goes to the condenser and the breaker points.

The secondary coil is made of many turns of fine wire. It is wound on top of the primary coil. One end of the secondary coil is connected to the primary coil and the other end is carried out of the coil at the high-tension terminal that connects to the distributor rotor.

8613. In the aircraft magneto system, if the P-lead is disconnected, the magneto will be

A—on regardless of ignition switch position.
B—grounded regardless of ignition switch position.
C—open regardless of ignition switch position.

If the P-lead is disconnected from a magneto, the magneto will be "hot," or ON, regardless of the position of the ignition switch.

Answers

8607 [C] (L02) AMT-P	8608 [B] (L02) AMT-P	8609 [A] (L02) AMT-P	8610 [B] (L02) AMT-P
8611 [C] (L02) AMT-P	8612 [B] (L02) AMT-P	8613 [A] (L02) AMT-P	

8614. (Refer to Figure 5.) Placing the engine master switch and battery switch to the on position and advancing the power lever, allows current to flow from the bus to the

A—fuel valve, external power receptacles, power lever switch, power lever relay coil, and ignition relay coil.
B—fuel valve, start switch, power lever switch, power lever relay coil, fuel pumps, and one side of the ignition relay contactor.
C—fuel valve, external power receptacles, undercurrent relay contacts, starter relay coil, and ignition relay contacts.

Closing the engine master switch by moving its contacts to their up position, the battery switch contacts to their down position and the contacts of the power lever switch to their up position, allows current to flow from the bus to the following:

- *The fuel valve through wire 2 and the upper contacts of the engine master switch*
- *The start switch through wire 2 and the lower contacts of the engine master switch*
- *The power lever relay coil through wire 2, the lower contacts of the engine master switch, wire 12, and the power lever switch. This causes all of the power lever relay contacts to move to their down position.*
- *The fuel pumps through wire 3 and the center contacts of the power lever relay*
- *One of the ignition solenoid contacts through wire 3 and the lower contacts of the power lever relay*

8615. (Refer to Figure 5.) With power applied to the bus bar, what switch changes will allow the ignition exciters test switch to function?

A—Engine master switch, battery switch, and power lever switch.
B—Engine master switch, start switch, and test switch.
C—Engine master switch and power lever switch.

When power is supplied to the bus, the ignition exciters test switch will function when the following switches are changed:

- *The engine master switch must be ON, in its up position.*

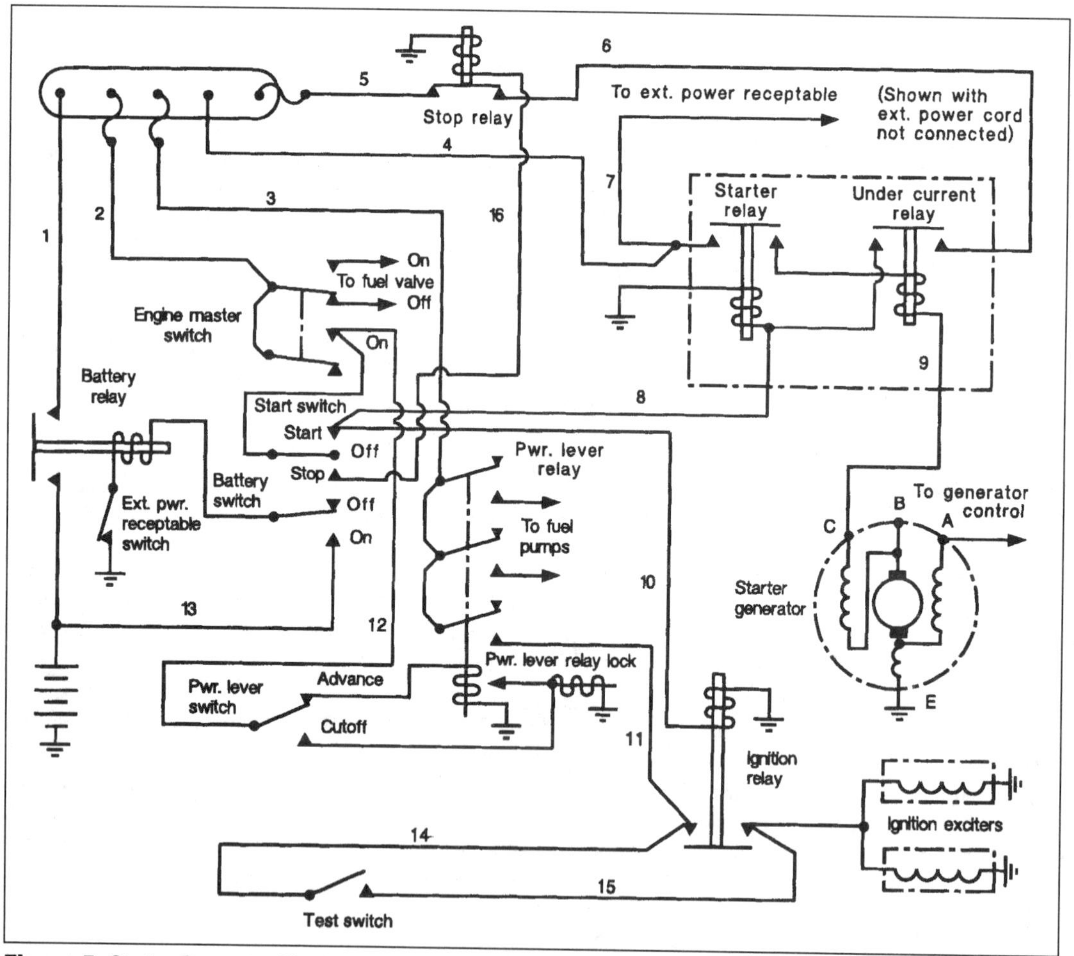

Figure 5. Starter-Generator Circuit

Answers
8614 [B] (L03) AMT-P 8615 [C] (L03) AMT-P

- *The power lever switch must be in its ADVANCE, or up, position so current can be supplied to the coil of the power lever relay to supply current to the left contact of the ignition relay.*
- *The test switch can now be closed to bypass the ignition relay and supply current to the ignition exciters.*

8616. (Refer to Figure 5.) The type of system depicted is capable of operating with

A—external power only.
B—either battery or external power.
C—battery power and external power simultaneously.

This electrical system can be operated by either the ship's battery or by an external power supply, but not by both at the same time. When the plug is inserted into the external power receptacle it opens the receptacle switch and stops current flow through the battery relay coil. The battery relay opens, disconnecting the battery from the system. This prevents both power sources being connected to the system at the same time.

8617. (Refer to Figure 5.) If wire No. 8 is broken or disconnected after starter rotation is initiated, and the power lever is advanced, the

A—starting sequence will continue normally.
B—starter will shut down, but the igniters will continue to fire.
C—starting sequence will discontinue.

If wire number 8 is broken or disconnected after the start sequence has begun, the starting sequence will continue normally.

Wire 8 supplies current to the starter relay coil, but when the starter relay contacts are closed, current flowing through the series starter coils in the starter generator causes the coil in the undercurrent relay to produce a strong enough field to close its contacts. Current from the bus flowing through the undercurrent relay contacts holds the starter relay coil energized, and the start sequence continues normally.

When the start switch is placed in the down, or STOP, position, or when the starter-generator requires too little current to hold the undercurrent relay closed, the starter relay contacts will open and discontinue the start sequence.

8618. (Refer to Figure 5.) When an external power source is connected to the aircraft,

A—the battery cannot be connected to the bus.
B—both battery power and external power are available to the bus.
C—the starter relay coil has a path to ground.

The way Figure 5 is drawn, both battery power and external power appear to be available to the bus at the same time. Figure 5 does not show what operates the external power receptacle switch, but this switch is automatically opened when the plug is inserted into the external power receptacle. When the external power receptacle switch is open, current cannot flow through the coil of the battery relay. This prevents both power sources being connected to the system at the same time.

8619. The purpose of an under current relay in a starter-generator system is to

A—provide a backup for the starter relay.
B—disconnect power from the starter-generator and ignition when sufficient engine speed is reached.
C—keep current flow to the starter-generator under the circuit capacity maximum.

The start procedure is begun by momentarily closing the start switch. Current flowing through this switch provides current to close the starter relay. When this relay closes, current flows through the series starter windings in the starter generator.

The high current required for starting the engine turning is enough to hold the undercurrent relay closed and supply current to the coil of the starter relay and the ignition relay.

When the engine starts, the current through the series starter winding decreases enough that the undercurrent relay will no longer remain closed. The starter and the ignition relay contacts will open, disconnecting current to the starter winding and de-energizing the ignition exciters.

8620. In a typical starter-generator system, under which of the following starting circumstances may it be necessary to use the start stop switch?

A—Hung start.
B—Hot start.
C—Contacts stick open.

The undercurrent relay will terminate the start procedure when the engine starts and the current through the series start windings in the starter generator drops to a predetermined low value.

The stop relay contacts are in series with the contacts of the undercurrent relay. When start switch is moved to the STOP position, it energizes the stop relay and opens the circuit to the starter relay coil and the ignition relay coil.

It may be necessary to use the start stop switch in the event of a hung start. A hung start is one in which ignition occurs but the RPM does not build up enough for the undercurrent relay to open.

Answers
8616 [B] (L03) AMT-P
8617 [A] (L03) AMT-P
8618 [A] (L03) AMT-P
8619 [B] (L03) AMT-P
8620 [A] (L03) AMT-P

8621. (Refer to Figure 5.) Which malfunctions will allow the igniters to operate when tested but be inoperative during a start attempt?

1. Conductor No. 10 broken.
2. Conductor No. 11 broken.
3. Ignition relay inoperative.
4. Conductor No. 12 broken.

A—1 or 4.
B—2 or 3.
C—1 or 3.

The igniters will operate when the test switch is closed but will fail to operate during a normal start attempt if there is an open in wire 10 which supplies current to the coil of the ignition relay, or if the ignition relay itself is inoperative.

8622. (Refer to Figure 5.) Which malfunctions will allow the igniters to operate normally during start but be inoperative when tested?

1. Conductor No. 14 broken.
2. Conductor No. 10 broken.
3. Conductor No. 15 broken.
4. Conductor No. 12 broken.

A—2 or 4.
B—1 or 3.
C—3 or 4.

The igniters will operate during a normal start, but will be inoperative when the test switch is closed if there is an open in either wire 14 or 15.

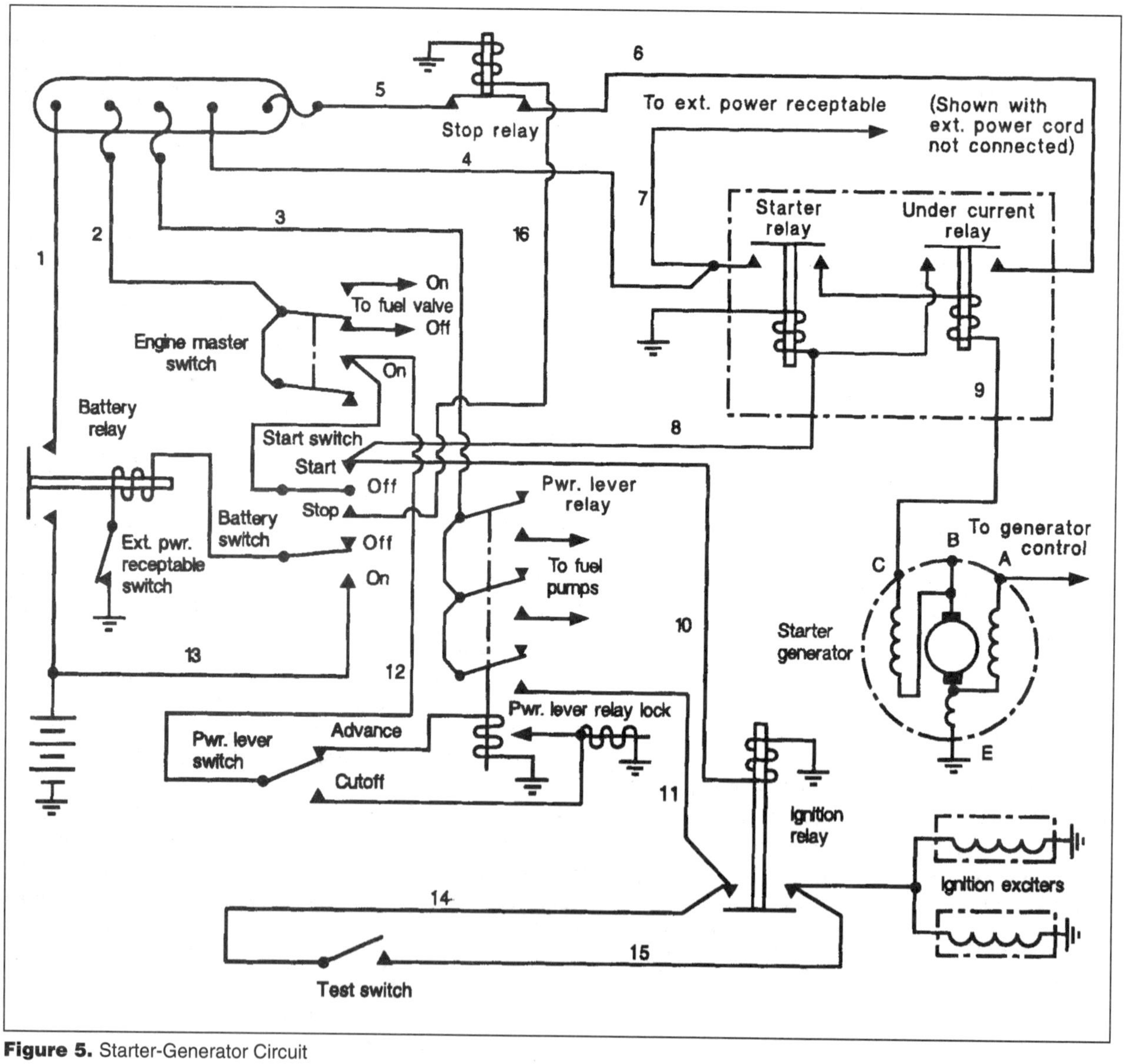

Figure 5. Starter-Generator Circuit

Answers
8621 [C] (L03) AMT-P
8622 [B] (L03) AMT-P

8623. When using an electric starter motor, current usage

A—is highest at the start of motor rotation.
B—remains relatively constant throughout the starting cycle.
C—is highest just before starter cutoff (at highest RPM).

Electric starter motors are series wound for their high starting torque. When the start switch is first closed, the current is very high, but as the motor begins to turn and produce a counter electromotive force, the current decreases.

8624. When using an electric starter motor, the current flow through it

A—is highest just before starter cutoff (at highest RPM.)
B—remains relatively constant throughout the starting cycle.
C—is highest at the start of motor rotation.

Electric starter motors are series wound for their high starting torque. When the start switch is first closed, the current flow is at it highest, and the battery voltage drops considerably. As the motor begins to turn and produce a counter electromotive force, the current decreases.

8625. The primary advantage of pneumatic (air turbine) starters over comparable electric starters for turbine engines is

A—a decreased fire hazard.
B—reduction gearing not required.
C—high power-to-weight ratio.

The primary advantage of an air turbine starter is its high power-to-weight ratio. An air turbine starter weighs approximately one fifth of that of an electric starter that produces an equivalent torque.

8626. A clicking sound heard at engine coast-down in a pneumatic starter incorporating a sprag clutch ratchet assembly is an indication of

A—gear tooth and/or pawl damage.
B—one or more broken pawl springs.
C—the pawls re-contacting and riding on the ratchet gear.

A pneumatic starter has a sprag clutch with a ratchet that automatically disengages the starter when the engine starts. Centrifugal force holds the pawls away from the ratchet while the engine is running.
When the engine is shut down and is coasting to a stop, the centrifugal force becomes too low to hold the pawls disengaged, and they will produce a clicking sound as they ride over the ratchet.

8627. Pneumatic starters are usually designed with what types of airflow impingement systems?

A—Radial inward flow turbine and axial-flow turbine.
B—Centrifugal compressor and axial-flow compressor.
C—Double entry centrifugal outward flow and axial-flow turbines.

Pneumatic starters may have an air impingement system of either the radial inflow turbine or the axial flow turbine type.

8628. Inspection of pneumatic starters by maintenance technicians usually includes checking the

A—oil level and magnetic drain plug condition.
B—stator and rotor blades for FOD.
C—rotor alignment.

Air turbine starters rotate at a speed of between 60,000 and 80,000 RPM, and lubrication provided by an integral oil supply is extremely important. Checking the oil level and the condition of the magnetic drain plug are important parts of routine starter maintenance.

8629. Air turbine starters are generally designed so that reduction gear distress or damage may be detected by

A—characteristic sounds from the starter assembly during engine start.
B—breakage of a shear section on the starter drive shaft.
C—inspection of a magnetic chip detector.

Air turbine starters have their own self-contained lubrication system with the oil held in the starter housing.
A magnetic chip detector is incorporated into the drain plug to warn the technician of any gear distress that produces flakes or shavings of metal.

8630. Airflow to the pneumatic starter from a ground unit is normally prevented from causing starter overspeed during engine start by

A—stator nozzle design that chokes airflow and stabilizes turbine wheel speed.
B—activation of a flyweight cutout switch.
C—a preset timed cutoff of the airflow at the source.

If the ground power unit supplies enough compressed air to the pneumatic starter to cause it to overspeed, a centrifugal cutout de-energizes the solenoid and the butterfly valve returns to its closed position, shutting off the air to the starter.

Answers

8623 [A] (L03) AMT-P	8624 [C] (L03) AMT-P	8625 [C] (L04) AMT-P	8626 [C] (L04) AMT-P
8627 [A] (L04) AMT-P	8628 [A] (L04) AMT-P	8629 [C] (L04) AMT-P	8630 [B] (L04) AMT-P

8631. A safety feature usually employed in pneumatic starters that is used if the clutch does not release from the engine drive at the proper time during start is the

A—flyweight cutout switch.
B—spring coupling release.
C—drive shaft shear point.

If the clutch in a pneumatic starter does not release when the engine starts, the engine will drive the starter to an excessively high speed. If this happens, a shear section in the drive shaft will break and prevent the engine driving the starter to destruction.

8632. A safety feature usually employed in pneumatic starters that is used to prevent the starter from reaching burst speed if inlet air does not terminate on schedule is the

A—drive shaft shear point.
B—stator nozzle design that chokes airflow and stabilizes turbine wheel speed.
C—spring coupling release.

If the inlet air driving a pneumatic starter does not terminate on schedule, the design of the inlet air passage will cause the nozzle to become choked and the turbine wheel will stabilize in an overspeed condition rather than continuing to accelerate and destroy itself.

8633. In the event a pneumatic start valve will not operate and the manual override must be used, the starter T-handle must be closed at scheduled starter drop out because

A—the starter will overheat.
B—the starter will overspeed at a given N^2.
C—the starter oil will be blown overboard.

When manually operating an air turbine starter by using the T-handle, it is important that the T-handle be turned to the CLOSED position when the N^2 reaches a specified RPM. If the valve is not closed at this time, the starter will overspeed.

8634. What factor is not used in the operation of an aircraft gas turbine engine fuel control unit?

A—Compressor inlet air temperature.
B—Mixture control position.
C—Power lever position.

The fuel-control unit on an aircraft gas turbine engine senses, among other parameters, compressor inlet air temperature and power lever position.

Gas turbine engines do not have a mixture control.

8635. In order to stabilize cams, springs, and linkages within the fuel control, manufacturers generally recommend that all final turbine engine trim adjustments be made in the

A—increase direction.
B—decrease direction.
C—decrease direction after over-adjustment.

When trimming a turbine engine fuel control, to stabilize the cams, springs, and linkages, all the final adjustments must be made in the increase direction.

8636. When trimming a turbine engine, the fuel control is adjusted to

A—produce as much power as the engine is capable of producing.
B—set idle RPM and maximum speed or EPR.
C—allow the engine to produce maximum RPM without regard to power output.

Trimming a turbine engine means adjusting the fuel control unit to set the idle RPM and the maximum speed or EPR of the engine.

8637. A supervisory electronic engine control (EEC) is a system that receives engine operating information and

A—adjusts a standard hydromechanical fuel control unit to obtain the most effective engine operation.
B—develops the commands to various actuators to control engine parameters.
C—controls engine operation according to ambient temperature, pressure, and humidity.

A supervisory electronic engine control (EEC) receives its engine operating data from sensors in the engine and controls a hydromechanical fuel control unit to obtain the most effective engine operation.

Answers

8631 [C] (L04) AMT-P
8632 [B] (L04) AMT-P
8633 [B] (L04) AMT-P
8634 [B] (M01) AMT-P
8635 [A] (M01) AMT-P
8636 [B] (M01) AMT-P
8637 [A] (M01) AMT-P

8638. A full-authority electronic engine control (EEC) is a system that receives all the necessary data for engine operation and

A—adjusts a standard hydromechanical fuel control unit to obtain the most effective engine operation.
B—develops the commands to various actuators to control engine parameters.
C—controls engine operation according to ambient temperature, pressure, and humidity.

A full-authority electronic engine control (EEC) performs all of the functions necessary to operate a turbofan engine most efficiently and safely in all modes of operation. This includes starting, accelerating, decelerating, takeoff, climb, cruise, and idle. It receives data from the aircraft and engine systems, provides data for the aircraft systems, and issues commands to the engine control actuators.

8639. In a supervisory EEC system, any fault in the EEC that adversely affects engine operation

A—causes redundant or backup units to take over and continue normal operation.
B—usually degrades performance to the extent that continued operation can cause damage to the engine.
C—causes an immediate reversion to control by the hydromechanical fuel control unit.

A safety feature of a supervisory ECC is that if any fault occurs which adversely affects the engine operation, the system will immediately revert the control to the hydromechanical control unit.

8640. The active clearance control (ACC) portion of an EEC system aids turbine engine efficiency by

A—adjusting stator vane position according to operating conditions and power requirements.
B—ensuring turbine blade to engine case clearances are kept to a minimum by controlling case temperatures.
C—automatically adjusting engine speed to maintain a desired EPR.

The active clearance control (ACC) portion of an ECC aids in maintaining peak engine efficiency by ensuring that the compressor and turbine blade clearances are kept to a minimum, thus reducing pressure losses due to leakage at the blade tips. This is done by the ACC directing cooling air through passages in the engine case to control engine case temperature, and thus the expansion of the case.

8641. What should be checked/changed to ensure the validity of a turbine engine performance check if an alternate fuel is to be used?

A—Fuel specific gravity setting.
B—Maximum RPM adjustment.
C—EPR gauge calibration.

Different turbine-engine fuels have different heat energy content. When making a run-up for a performance check, the fuel specific gravity setting on the fuel control must be changed to give validity to the check.

8642. The generally acceptable way to obtain accurate on-site temperature prior to performing engine trimming is to

A—call the control tower to obtain field temperature.
B—observe the reading on the aircraft Outside Air Temperature (OAT) gauge.
C—hang a thermometer in the shade of the nose wheel-well until the temperature reading stabilizes.

Accurate temperature is critical when trimming a turbine engine. To determine the temperature at the engine inlet, it is a good policy to hang a thermometer in the shade of the nose wheel-well until its temperature reading stabilizes. Use this as the actual ambient temperature.

8643. An aircraft should be facing into the wind when trimming an engine. However, if the velocity of the wind blowing into the intake is excessive, it is likely to cause a

A—false low exhaust gas temperature reading.
B—trim setting resulting in engine overspeed.
C—false high compression and turbine discharge pressure, and a subsequent low trim.

If the wind blowing into the engine inlet while trimming the fuel control is excessive, the compression and turbine discharge pressures will be too high. When the engine is operated with the normal amount of wind flowing into it, the trimmed conditions will be too low.

8644. Generally, the practice when trimming an engine is to

A—turn all accessory bleed air off.
B—turn all accessory bleed air on.
C—make adjustments (as necessary) for all engines on the same aircraft with accessory bleed air settings the same—either on or off.

When trimming a turbine engine, all of the overboard air valves should be closed and all of the accessory air bleed must be turned off.

Answers

8638 [B] (M01) AMT-P	8639 [C] (M01) AMT-P	8640 [B] (M01) AMT-P	8641 [A] (M01) AMT-P
8642 [C] (M01) AMT-P	8643 [C] (M01) AMT-P	8644 [A] (M01) AMT-P	

8645. A reciprocating engine automatic mixture control responds to changes in air density caused by changes in

A—altitude or humidity.
B—altitude only.
C—altitude or temperature.

An automatic mixture control in a pressure carburetor or fuel injection system responds to changes in air density that are caused by changes in both altitude and temperature.

As either altitude or temperature increases, the air becomes less dense, and the AMC decreases the amount of fuel metered to the engine to keep the fuel-air mixture ratio constant.

8646. On a float-type carburetor, the purpose of the economizer valve is to

A—provide extra fuel for sudden acceleration of the engine.
B—maintain the leanest mixture possible during cruising best power.
C—provide a richer mixture and cooling at maximum power output.

An economizer system, also called a power enrichment system, provides additional fuel at full power operation to remove some of the excess heat from the cylinders.

This system is called an economizer system because it allows an economical operation at conditions other than full power.

8647. The fuel metering force of a conventional float-type carburetor in its normal operating range is the difference between the pressure acting on the discharge nozzle located within the venturi and the pressure

A—acting on the fuel in the float chamber.
B—of the fuel as it enters the carburetor.
C—of the air as it enters the venturi (impact pressure).

The fuel-metering force in a float-type carburetor is the pressure difference between the pressure in the float bowl and the pressure at the discharge nozzle which is located in the venturi.

The low pressure at the discharge nozzle is determined by the amount of air flowing into the engine through the venturi.

8648. If the main air bleed of a float-type carburetor becomes clogged, the engine will run

A—lean at rated power.
B—rich at rated power.
C—rich at idling.

A clogged main air bleed will cause the mixture to be too rich at the rated power of the engine.

It will have no effect on the mixture at idling because the idling system does not use the main air bleed.

8649. Which method is commonly used to adjust the level of a float in a float-type carburetor?

A—Lengthening or shortening the float shaft.
B—Add or remove shims under the needle-valve seat.
C—Change the angle of the float arm pivot.

A widely used method of adjusting the float level on a float-type carburetor is by adding or removing shims under the needle-valve seat.

8650. What is the possible cause of an engine running rich at full throttle if it is equipped with a float-type carburetor?

A—Float level too low.
B—Clogged main air bleed.
C—Clogged atmospheric vent.

A clogged main air bleed will cause an engine to run too rich at full throttle.

This principle of restricting the main air bleed at full throttle operation is used in some carburetors as a power-enrichment system.

8651. One of the things a metering orifice in a main air bleed helps to accomplish (at a given altitude) in a carburetor is

A—pressure in the float chamber to increase as airflow through the carburetor increases.
B—a progressively richer mixture as airflow through the carburetor increases.
C—better fuel vaporization and control of fuel discharge, especially at lower engine speeds.

A metering orifice in the main air bleed of a float-type carburetor meters the correct amount of air into the fuel just before it reaches the main discharge nozzle. This air emulsifies the fuel to decrease its density and make it easier to vaporize, especially at the lower engine speeds.

Answers
8645 [C] (M02) AMT-P
8646 [C] (M02) AMT-P
8647 [A] (M02) AMT-P
8648 [B] (M02) AMT-P
8649 [B] (M02) AMT-P
8650 [B] (M02) AMT-P
8651 [C] (M02) AMT-P

8652. A punctured float in a float-type carburetor will cause the fuel level to

A—lower, and enrich the mixture.
B—rise, and enrich the mixture.
C—rise, and lean the mixture.

A punctured float, in a float-type carburetor, will allow fuel to enter the float, making it less buoyant. The fuel will have to rise to a higher level in the float bowl to shut the fuel off.

This higher level of fuel in the float bowl will enrich the mixture.

8653. The back-suction mixture control system operates by

A—varying the pressure within the venturi section.
B—varying the pressure acting on the fuel in the float chamber.
C—changing the effective cross-sectional area of the main metering orifice (jet).

The back-suction mixture control changes the fuel-air mixture ratio by varying the pressure on the top of the fuel in the float bowl.

Lowering the pressure on this fuel leans the mixture.

8654. If an aircraft engine is equipped with a carburetor that is not compensated for altitude and temperature variations, the fuel/air mixture will become

A—leaner as either the altitude or temperature increases.
B—richer as the altitude increases and leaner as the temperature increases.
C—richer as either the altitude or temperature increases.

A noncompensated carburetor (a carburetor that cannot be adjusted for mixture changes with altitude) will produce a richer mixture as the air density decreases because of an increase in either altitude or temperature.

The reason for this enrichment is that the carburetor meters the fuel according to the volume of air, but the engine burns it according to its weight.

The weight of a given volume of air decreases as both altitude and temperature increase, so the mixture becomes richer.

8655. Float-type carburetors which are equipped with economizers are normally set for

A—their richest mixture delivery and leaned by means of the economizer system.
B—the economizer system to supplement the main system supply at all engine speeds above idling.
C—their leanest practical mixture delivery at cruising speeds and enriched by means of the economizer system at higher power settings.

A carburetor equipped with an economizer system is adjusted so the mixture will be sufficiently lean for economical operation under all conditions other than full power.

Under full power conditions, the economizer system adds fuel to enrich the mixture enough that detonation will not occur.

8656. If a float-type carburetor becomes flooded, the condition is most likely caused by

A—a leaking needle valve and seat assembly.
B—the accelerating pump shaft being stuck.
C—a clogged back-suction line.

If a needle valve and seat assembly in a float-type carburetor leaks, the fuel level will rise above the discharge level of the main discharge nozzle and the carburetor will become flooded.

8657. If an engine is equipped with a float-type carburetor and the engine runs excessively rich at full throttle, a possible cause of the trouble is a clogged

A—main air bleed.
B—back-suction line.
C—atmospheric vent line.

Under full-throttle operation, the main air bleed must introduce enough air into the fuel to prevent the mixture becoming excessively rich.

If this air bleed becomes restricted, or clogged, the engine will run excessively rich at wide-open throttle.

The power enrichment system used on some float-type carburetors enriches the mixture at full power by restricting the main air bleed.

8658. What occurs when a back-suction type mixture control is placed in IDLE CUTOFF?

A—The fuel passages to the main and idle jets will be closed by a valve.
B—The float chamber will be vented to a negative pressure area.
C—The fuel passage to the idle jet will be closed by a valve.

The back-suction mixture control regulates the fuel-air mixture by varying the pressure above the fuel in the float chamber.

When this type of carburetor is placed in the IDLE CUTOFF position, the float chamber is vented to a negative pressure.

Answers

8652 [B] (M02) AMT-P	8653 [B] (M02) AMT-P	8654 [C] (M02) AMT-P	8655 [C] (M02) AMT-P
8656 [A] (M02) AMT-P	8657 [A] (M02) AMT-P	8658 [B] (M02) AMT-P	

8659. Which of the following best describes the function of an altitude mixture control?

A—Regulates the richness of the fuel/air charge entering the engine.
B—Regulates the air pressure above the fuel in the float chamber.
C—Regulates the air pressure in the venturi.

An altitude mixture control allows the pilot to maintain a fuel-air mixture ratio entering the engine that will produce the desired engine operation.

This is the same as regulating the richness of the fuel-air charge entering the engine.

8660. Select the correct statement concerning the idle system of a conventional float-type carburetor.

A—The low-pressure area created in the throat of the venturi pulls the fuel from the idle passage.
B—Climatic conditions have very little effect on idle mixture requirements.
C—The low pressure between the edges of the throttle valve and the throttle body pulls the fuel from the idle passage.

In a conventional float-type carburetor at idle, there is not enough air flowing through the venturi to pull fuel from the main discharge nozzle.

The idle discharge nozzle is located in the throttle body at the edge of the throttle valve, and all of the air flowing into the engine for idling must pass between the edges of the throttle valve and the throttle body. It produces a low pressure that pulls the fuel-air emulsion from the idle passage.

8661. On an engine equipped with a pressure-type carburetor, fuel supply in the idling range is ensured by the inclusion in the carburetor of

A—a spring in the unmetered fuel chamber to supplement the action of normal metering forces.
B—an idle metering jet that bypasses the carburetor in the idle range.
C—a separate boost venturi that is sensitive to the reduced airflow at start and idle speeds.

In a pressure carburetor, there is a spring in the unmetered fuel chamber that holds the poppet valve slightly off its seat to ensure a uniform flow of fuel under conditions of low airflow.

8662. The economizer system of a float-type carburetor performs which of the following functions?

A—It supplies and regulates the fuel required for all engine speeds.
B—It supplies and regulates the additional fuel required for all engine speeds above cruising.
C—It regulates the fuel required for all engine speeds and all altitudes.

The economizer system on a float-type carburetor adds additional fuel to that which is metered through the main metering jet at all conditions above cruise.

It is for this reason that economizer systems are often called power enrichment systems.

8663. How will the mixture of an engine be affected if the bellows of the automatic mixture control (AMC) in a pressure carburetor ruptures while the engine is operating at altitude?

A—It will become leaner.
B—No change will occur until the altitude changes.
C—It will become richer.

In a pressure-injection carburetor, air from the impact tubes flows past the automatic mixture control (AMC) needle valve, into chamber A of the regulator and out through mixture control bleeds into chamber B.

Pressure of an inert gas inside the AMC bellows is balanced by a spring force that tries to collapse the bellows and open the passage.

As the aircraft goes up in altitude, atmospheric pressure on the outside of the bellows decreases. The pressure of the gas inside the AMC overcomes the spring pressure enough to close the needle valve and restrict the flow of air. This causes the air metering force to lean the mixture progressively as altitude increases.

If the bellows ruptures and allows the inert gas to escape from the AMC unit, the spring will pull the needle back, opening the passage and allowing a full flow of impact air into chamber A. This will increase the air metering force and open the poppet valve, allowing the mixture to become richer.

8664. The fuel level within the float chamber of a properly adjusted float-type carburetor will be

A—slightly higher than the discharge nozzle outlet.
B—slightly lower than the discharge nozzle outlet.
C—at the same level as the discharge nozzle outlet.

The fuel level of a properly adjusted float-type carburetor will be slightly lower than the discharge nozzle outlet. This prevents fuel flowing out of the discharge nozzle when the aircraft is sitting in a level-ground attitude.

Answers

8659 [A] (M02) AMT-P
8660 [C] (M02) AMT-P
8661 [A] (M02) AMT-P
8662 [B] (M02) AMT-P
8663 [C] (M02) AMT-P
8664 [B] (M02) AMT-P

8665. The metered fuel pressure (chamber C) in an injection-type carburetor

A—is held constant throughout the entire engine operating range.
B—varies according to the position of the poppet valve located between chamber D (unmetered fuel) and chamber E (engine-driven fuel pump pressure).
C—will be approximately equal to the pressure in chamber A (impact pressure).

Metered fuel pressure in chamber C of a pressure-injection carburetor is held constant by a constant-pressure discharge valve. This pressure acts as a reference pressure.

The total metering force is made up of the air metering force across the air diaphragm between chambers A and B and the fuel metering force across the fuel diaphragm between chambers C and D.

The fuel metering force is a balance between the unmetered, but regulated, fuel pressure in chamber D and the metered fuel pressure in chamber C, which is held constant throughout the engine operating range.

8666. Select the statement which is correct relating to a fuel level check of a float-type carburetor.

A—Use 5 pounds fuel pressure for the test if the carburetor is to be used in a gravity fuel feed system.
B—Block off the main and idle jets to prevent a continuous flow of fuel through the jets.
C—Do not measure the level at the edge of the float chamber.

The fuel level must not be measured at the edge of the float chamber because of the tendency of the fuel to wet the walls of the chamber. At the wall, the fuel will be slightly higher than it is in the center.

A straightedge should be placed across the parting surface of the float bowl and the level measured away from the edge and away from the float.

8667. What carburetor component measures the amount of air delivered to the engine?

A—Economizer valve.
B—Automatic mixture control.
C—Venturi.

A venturi is used to measure the amount of the air pulled into the cylinders of an engine.

The pressure drop in the throat of a venturi is directly proportional to the velocity of the air flowing through it.

8668. If a float-type carburetor leaks fuel when the engine is stopped, a likely cause is that the

A—float needle valve is worn or otherwise not seated properly.
B—float level is adjusted too low.
C—main air bleed is clogged.

A float-type carburetor leaking when the engine is stopped is caused by fuel rising in the float bowl to a level higher than the tip of the fuel discharge nozzle.

A leaking needle valve is the most likely cause of the fuel level rising when the engine is not running.

8669. Fuel is discharged for idling speeds on a float-type carburetor

A—from the idle discharge nozzle.
B—in the venturi.
C—through the idle discharge air bleed.

In a float-type carburetor at idle speed, fuel is discharged from the idle discharge nozzle which is located opposite the throttle valve in the carburetor throat.

8670. When air passes through the venturi of a carburetor, what three changes occur?

A—Velocity increases, temperature increases, and pressure decreases.
B—Velocity decreases, temperature increases, and pressure increases.
C—Velocity increases, temperature decreases, and pressure decreases.

When air passes through a venturi, its velocity increases, and both its temperature and pressure decrease.

8671. Where is the throttle valve located on a float-type carburetor?

A—Between the venturi and the discharge nozzle.
B—After the main discharge nozzle and venturi.
C—After the venturi and just before the main discharge nozzle.

The throttle valve in a float-type carburetor is located after (downstream of) the main discharge nozzle and the venturi.

The discharge nozzle is located in the venturi.

Answers
8665 [A] (M02) AMT-P
8666 [C] (M02) AMT-P
8667 [C] (M02) AMT-P
8668 [A] (M02) AMT-P
8669 [A] (M02) AMT-P
8670 [C] (M02) AMT-P
8671 [B] (M02) AMT-P

8672. An aircraft carburetor is equipped with a mixture control in order to prevent the mixture from becoming too

A—lean at high altitudes.
B—rich at high altitudes.
C—rich at high speeds.

An aircraft carburetor is equipped with a mixture control to prevent the fuel-air mixture from becoming too rich as the aircraft goes up in altitude and the air becomes less dense.

8673. Which of the following is NOT a function of the carburetor venturi?

A—Proportions the fuel/air mixture.
B—Regulates the idle system.
C—Limits the airflow at full throttle.

The carburetor venturi has nothing to do with regulating the idle system of the carburetor.

The venturi does proportion the fuel/air mixture and limits the airflow at full throttle.

8674. Idle cutoff is accomplished on a carburetor equipped with a back-suction mixture control by

A—introducing low pressure (intake manifold) air into the float chamber.
B—turning the fuel selector valve to OFF.
C—the positive closing of a needle and seat.

Idle cutoff in a carburetor equipped with a back-suction mixture control is accomplished by introducing low-pressure air (suction) into the float chamber.

This low pressure lowers the level of fuel in the discharge nozzle so much that the low pressure caused by air flowing through the venturi cannot pull fuel from it.

8675. One purpose of an air bleed in a float-type carburetor is to

A—increase fuel flow at altitude.
B—meter air to adjust the mixture.
C—decrease fuel density and destroy surface tension.

An air bleed in a float-type carburetor mixes air in the fuel between the metering jet and the discharge nozzle. This decreases the density of the fuel and destroys its surface tension, resulting in better vaporization and control of fuel discharge, especially at lower engine speeds.

8676. To determine the float level in a float-type carburetor, a measurement is usually made from the top of the fuel in the float chamber to the

A—parting surface of the carburetor.
B—top of the float.
C—centerline of the main discharge nozzle.

The fuel level in the float bowl of a float-type carburetor is measured between the parting surface of the carburetor float bowl and the surface of the fuel in the bowl measured at a point away from both the float chamber wall and the float.

8677. The throttle valve of float-type aircraft carburetors is located

A—ahead of the venturi and main discharge nozzle.
B—after the main discharge nozzle and ahead of the venturi.
C—between the venturi and the engine.

The throttle valve in a float-type carburetor is a butterfly-type valve located between the venturi and the engine.

8678. Why must a float-type carburetor supply a rich mixture during idle?

A—Engine operation at idle results in higher than normal volumetric efficiency.
B—Because at idling speeds the engine may not have enough airflow around the cylinders to provide proper cooling.
C—Because of reduced mechanical efficiency during idle.

The idle mixture is considerably richer than a normal mixture because at idling speeds the mixture distribution is sometimes uneven, and the engine may not have enough airflow around the cylinders to provide proper cooling.

8679. What component is used to ensure fuel delivery during periods of rapid engine acceleration?

A—Acceleration pump.
B—Water injection pump.
C—Power enrichment unit.

An accelerator system usually contains an accelerator pump to provide a momentarily rich mixture during periods of rapid engine acceleration.

Under these conditions, the normal metering system is unable to provide an adequately rich mixture.

8680. The device that controls the ratio of the fuel/air mixture to the cylinders is called a

A—throttle valve.
B—mixture control.
C—metering jet.

Most aircraft reciprocating engines are equipped with a mixture control that allows the pilot to control the fuel-air mixture ratio as it enters the cylinders of the engine.

Answers

8672 [B] (M02) AMT-P	8673 [B] (M02) AMT-P	8674 [A] (M02) AMT-P	8675 [C] (M02) AMT-P
8676 [A] (M02) AMT-P	8677 [C] (M02) AMT-P	8678 [B] (M02) AMT-P	8679 [A] (M02) AMT-P
			8680 [B] (M02) AMT-P

8681. The device that controls the volume of the fuel/air mixture to the cylinders is called a

A—mixture control.
B—metering jet.
C—throttle valve.

The throttle valve in a carburetor determines the volume of fuel-air mixture allowed to enter the cylinders.

8682. Which statement is correct regarding a continuous-flow fuel injection system used on many reciprocating engines?

A—Fuel is injected directly into each cylinder.
B—Fuel is injected at each cylinder intake port.
C—Two injector nozzles are used in the injector fuel system for various speeds.

A continuous-flow fuel injection system delivers a continuous flow of metered fuel to the intake port of each cylinder and not into the combustion chambers.

8683. During the operation of an aircraft engine, the pressure drop in the carburetor venturi depends primarily upon the

A—air temperature.
B—barometric pressure.
C—air velocity.

The pressure drop in the venturi of a carburetor is a function of the velocity of the air flowing into the engine.

The relationship between pressure and velocity in a stream of moving fluid is explained by Bernoulli's principle.

8684. Which of the following causes a single diaphragm accelerator pump to discharge fuel?

A—An increase in venturi suction when the throttle valve is open.
B—An increase in manifold pressure that occurs when the throttle valve is opened.
C—A decrease in manifold pressure that occurs when the throttle valve is opened.

When the throttle is suddenly opened, there is an increase in manifold pressure (less suction).

The lack of suction on the diaphragm allows the spring to push the diaphragm over and force the fuel in the accelerator pump chamber out into the airstream.

8685. At what engine speed does the main metering jet actually function as a metering jet in a float-type carburetor?

A—All RPM's.
B—Cruising RPM only.
C—All RPM's above idle range.

The main metering jet in a float-type carburetor functions as a metering jet at all RPMs above the idling range.

In the idling range, the idling metering jet meters the fuel.

8686. An aircraft engine continuous cylinder fuel injection system normally discharges fuel during which stroke(s)?

A—Intake.
B—Intake and compression.
C—All (continuously).

A continuous-flow, fuel injection system discharges fuel during all of the strokes. This is the reason this type of system is called a continuous-flow system.

8687. What is the purpose of the carburetor accelerating system?

A—Supply and regulate the fuel required for engine speeds above idle.
B—Temporarily enrich the mixture when the throttle is suddenly opened.
C—Supply and regulate additional fuel required for engine speeds above cruising.

There is normally a temporary lag between the time the throttle is opened and the time the engine builds up its speed.

This lag is due to the time required to transition between the idle system and the main metering system.

To prevent this lag, most carburetors are equipped with an accelerating system that provides a momentarily rich mixture during this transition time.

8688. When troubleshooting an engine for too rich a mixture to allow the engine to idle, what would be a possible cause?

A—A primer line open.
B—Mixture setting too rich.
C—Air leak in the intake manifold.

If the engine will not idle because the mixture is too rich, and the idle mixture adjustment on the carburetor will not lean it sufficiently, there is a possibility that the primer line is open, allowing the primer to feed fuel into the engine.

Answers

8681 [C] (M02) AMT-P	8682 [B] (M03) AMT-P	8683 [C] (M03) AMT-P	8684 [B] (M03) AMT-P
8685 [C] (M03) AMT-P	8686 [C] (M03) AMT-P	8687 [B] (M03) AMT-P	8688 [A] (M03) AMT-P

8689. What is the relationship between the accelerating pump and the enrichment valve in a pressure injection carburetor?

A—No relationship since they operate independently.
B—Fuel pressure affects both units.
C—The accelerating pump actuates the enrichment valve.

There are different types of pressure-injection carburetors that use different types of enrichment and accelerating systems.

The majority of pressure-injection carburetors use a balance between metered and unmetered fuel pressure to open the enrichment valve, and a balance between manifold pressure and spring force to actuate the accelerating pump.

The enrichment valve produces an increase of steady-state flow of fuel, while the accelerating pump provides a momentary increase in the fuel flow when the throttle is suddenly opened.

8690. What is the relationship between the pressure existing within the throat of a venturi and the velocity of the air passing through the venturi?

A—There is no direct relationship between the pressure and the velocity.
B—The pressure is directly proportional to the velocity.
C—The pressure is inversely proportional to the velocity.

The pressure in the throat of a venturi is inversely proportional to the velocity of the air flowing through it.

This is in accordance with Bernoulli's principle.

8691. Which of the following is least likely to occur during operation of an engine equipped with a direct cylinder fuel injection system?

A—Afterfiring.
B—Kickback during start.
C—Backfiring.

Backfiring, which is the ignition of the fuel-air mixture in the induction system of an engine during the time of valve overlap, is least likely to occur on an engine equipped with direct cylinder fuel injection. This is because there is no combustible fuel-air mixture in the induction system. The fuel is injected directly into the combustion chamber of the cylinder.

A direct cylinder fuel injection system is not to be confused with a continuous-flow fuel injection system.

8692. What carburetor component actually limits the desired maximum airflow to the engine at full throttle?

A—Throttle valve.
B—Venturi.
C—Manifold intake.

The venturi limits the maximum airflow into the engine at full throttle operation.

At conditions other than full throttle, the position of the butterfly-type throttle valve limits the amount of air.

8693. On a carburetor without an automatic mixture control as you ascend to altitude, the mixture will

A—be enriched.
B—be leaned.
C—not be affected.

If the carburetor does not have an automatic mixture control, the mixture ratio will become richer as the aircraft goes up in altitude and the air density becomes less.

8694. During engine operation, if carburetor heat is applied, it will

A—increase fuel/air ratio.
B—increase engine RPM.
C—decrease the air density to the carburetor.

When carburetor heat is applied, the temperature of the air entering the engine is increased. Increasing its temperature decreases the density of the intake air.

8695. The desired engine idle speed and mixture setting

A—is adjusted with engine warmed up and operating.
B—should give minimum RPM with maximum manifold pressure.
C—is usually adjusted in the following sequence; speed first, then mixture.

Engine idle speed and mixture setting can be adjusted only when the engine is warmed up and operating properly.

Answers

8689 [A] (M03) AMT-P	8690 [C] (M03) AMT-P	8691 [C] (M03) AMT-P	8692 [B] (M03) AMT-P
8693 [A] (M03) AMT-P	8694 [C] (M03) AMT-P	8695 [A] (M03) AMT-P	

8696. A nine-cylinder radial engine, using a multiple-point priming system with a central spider, will prime which cylinders?

A—One, two, three, eight, and nine.
B—All cylinders.
C—One, three, five, and seven.

A radial engine, using multiple-point priming, primes all of the cylinders above the center line of the engine.

In a nine-cylinder engine, these are cylinders 1, 2, 3, 8, and 9.

8697. What is a function of the idling air bleed in a float-type carburetor?

A—It provides a means for adjusting the mixture at idle speeds.
B—It vaporizes the fuel at idling speeds.
C—It aids in emulsifying/vaporizing the fuel at idle speeds.

The idle air bleed on a float-type carburetor aids in emulsifying the fuel drawn from the idle system during idle-speed operation.

8698. If the volume of air passing through a carburetor venturi is reduced, the pressure at the venturi throat will

A—decrease.
B—be equal to the pressure at the venturi outlet.
C—increase.

When the volume of air passing through the carburetor venturi is reduced by partially closing the throttle valve, the pressure at the venturi throat will increase.

8699. (Refer to Figure 6.) Which curve most nearly represents an aircraft engine's fuel/air ratio throughout its operating range?

A—1.
B—3.
C—2.

Curve 1 is not correct, because the mixture does not get progressively richer as the RPM increases.

Curve 2 is correct, because a richer mixture is provided for idling when scavenging is poor and also for full power operation when extra fuel is needed for cooling.

Curve 3 is not correct, because the actual mixture ratio needs are exactly opposite to that shown in this curve.

Curve 4 is not correct, because the mixture does not get progressively richer as the RPM increases.

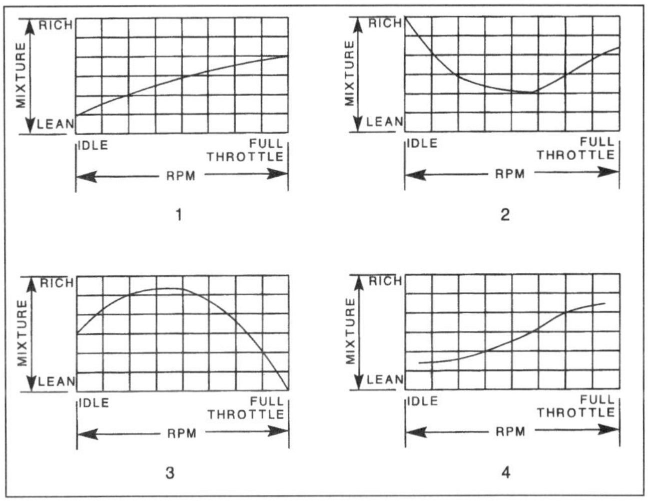

Figure 6

8700. What will occur if the vapor vent float in a pressure carburetor loses its buoyancy?

A—The amount of fuel returning to the fuel tank from the carburetor will be increased.
B—The engine will continue to run after the mixture control is placed in IDLE CUTOFF.
C—A rich mixture will occur at all engine speeds.

The vapor vent floats in a pressure-injection carburetor ride on top of the fuel in chambers D and E.

When vapors collect in these chambers, the float drops down and opens the needle valve so fuel and vapors can return to the fuel tank.

If either float loses its buoyancy, the needle valve will remain off its seat and there will be a continuous flow of fuel back to the tank.

This return of fuel does not affect either the engine operation or the fuel consumption, but it does cause the flow indication to be incorrect.

In modern pressure carburetors, there are no floats or valves in this system, but a calibrated amount of fuel continually flows back into one of the tanks.

Answers
8696 [A] (M03) AMT-P
8697 [C] (M03) AMT-P
8698 [C] (M03) AMT-P
8699 [C] (M04) AMT-P
8700 [A] (M04) AMT-P

8701. What method is ordinarily used to make idle speed adjustments on a float-type carburetor?

A—An adjustable throttle stop or linkage.
B—An orifice and adjustable tapered needle.
C—An adjustable needle in the drilled passageway which connects the airspace of the float chamber and the carburetor venturi.

Idle speed adjustment on a float-type carburetor is made by adjusting the amount the throttle valve remains away from the throttle body wall when the throttle control is against its stop.

This adjustment is normally made with an adjustable throttle stop screw.

8702. For what primary purpose is a turbine engine fuel control unit trimmed?

A—To obtain maximum thrust output when desired.
B—To properly position the power levers.
C—To adjust the idle RPM.

The adjustments allowed to be made on a turbine engine are the adjustment of the idling RPM and the maximum RPM adjustment. These adjustments are commonly called trimming the engine.

An engine is trimmed to allow it to obtain maximum thrust output when it is desired.

8703. Which type of fuel control is used on most of today's turbine engines?

A—Electromechanical.
B—Mechanical.
C—Hydromechanical or electronic.

Most of the modern fuel controls used on turbine engines are of the hydromechanical or electronic type.

8704. Under which of the following conditions will the trimming of a turbine engine be most accurate?

A—High wind and high moisture.
B—High moisture and low wind.
C—No wind and low moisture.

To accurately trim a turbine engine, which is the adjustment of the fuel control unit, there should be no wind to create an untrue pressure at the engine intake.

There should also be a low moisture content in the air. This is because moisture, or water vapor, is much less dense than standard (dry) air.

8705. (1) The mixture used at rated power in air cooled reciprocating engines is richer than the mixture used through the normal cruising range.

(2) The mixture used at idle in air cooled reciprocating engines is richer than the mixture used at rated power.

Regarding the above statements,

A—only No. 1 is true.
B—only No. 2 is true.
C—both No. 1 and No. 2 are true.

Statement (1) is true. The mixture ratio used when an engine is developing its rated horsepower is richer than the ratio used throughout the normal cruising range. The additional fuel is used for cooling.

Statement (2) is also true. The mixture used at idling is richer than the mixture used at rated power because incomplete scavenging and uneven mixture distribution during idle operation requires an excessively rich mixture.

8706. Under which of the following conditions would an engine run lean even though there is a normal amount of fuel present?

A—The use of too high an octane rating fuel.
B—Incomplete fuel vaporization.
C—The carburetor air heater valve in the HOT position.

Incomplete fuel vaporization causes an excess of air for the amount of vaporized fuel.

Incomplete fuel vaporization causes an engine to operate lean even though a normal amount of fuel has been metered.

8707. During idle mixture adjustments, which of the following is normally observed to determine when the correct mixture has been achieved?

A—Changes in fuel/air pressure ratio.
B—Fuel flowmeter.
C—Changes in RPM or manifold pressure.

The correct idling mixture adjustments are determined by watching the changes in the RPM and the manifold pressure as the mixture control is slowly moved into the IDLE CUTOFF position.

Answers

8701 [A] (M04) AMT-P	8702 [A] (M04) AMT-P	8703 [C] (M04) AMT-P	8704 [C] (M04) AMT-P
8705 [C] (M04) AMT-P	8706 [B] (M04) AMT-P	8707 [C] (M04) AMT-P	

8708. An indication that the optimum idle mixture has been obtained occurs when the mixture control is moved to IDLE CUTOFF and manifold pressure

A—decreases momentarily and RPM drops slightly before the engine ceases to fire.
B—increases momentarily and RPM drops slightly before the engine ceases to fire.
C—decreases and RPM increases momentarily before the engine ceases to fire.

The optimum idle mixture setting causes a slight decrease in the manifold pressure and a slight rise in RPM as the mixture control is pulled back into the IDLE CUTOFF position. These changes occur just before the engine stops firing.

8709. The use of less than normal throttle opening during starting will cause

A—a rich mixture.
B—a lean mixture.
C—backfire due to lean fuel/air ratio.

If the throttle is opened less than normal during the engine starting procedure, a rich mixture will be drawn into the cylinders.

8710. When checking the idle mixture on a carburetor, the engine should be idling normally, then pull the mixture control toward the IDLE CUTOFF position. A correct idling mixture will be indicated by

A—an immediate decrease in RPM.
B—a decrease of 20 to 30 RPM before quitting.
C—an increase of 10 to 50 RPM before decreasing.

The correct idling fuel-air mixture is indicated when there is a slight rise in RPM before the engine stops running as the mixture control is slowly moved into the IDLE CUTOFF position.

8711. When a new carburetor is installed on an engine,

A—warm up the engine and adjust the float level.
B—do not adjust the idle mixture setting; this was accomplished on the flow bench.
C—and the engine is warmed up to normal temperatures, adjust the idle mixture, then the idle speed.

After a new carburetor is installed on an aircraft engine, the engine must be warmed up to its normal operating temperature and the idle mixture and idle speed adjustments made.

8712. The purpose of the back-suction mixture control in a float-type carburetor is to adjust the mixture by

A—regulating the pressure drop at the venturi.
B—regulating the pressure on the fuel in the float chamber.
C—regulating the suction on the mixture from behind the throttle valve.

A back-suction mixture control in a float-type carburetor adjusts the fuel-air mixture ratio by regulating the air pressure on the fuel in the float bowl.

This pressure determines the fuel level at the discharge nozzle.

8713. Reciprocating engine power will be decreased at all altitudes if the

A—air density is increased.
B—humidity is increased.
C—manifold pressure is increased.

An increase in humidity decreases the power produced by an engine, because the amount of power released from the fuel is determined by the mass of air mixed with the fuel when it is burned.

Water vapor is less dense than dry air, so humid air has less mass than an equal volume of dry air.

8714. If the idling jet becomes clogged in a float-type carburetor, the

A—engine operation will not be affected at any RPM.
B—engine will not idle.
C—idle mixture becomes richer.

A float-type carburetor with a clogged idling jet will not provide fuel to the engine when the throttle is closed to the idle position.

The engine will not run at idle speeds.

8715. An aircraft engine equipped with a pressure-type carburetor is started with the

A—primer while the mixture control is positioned at IDLE CUTOFF.
B—mixture control in the FULL-RICH position.
C—primer while the mixture control is positioned at the FULL-LEAN position.

An engine equipped with a pressure carburetor is started with fuel supplied to the cylinders by the priming system.

The engine is started with the mixture control in the IDLE CUTOFF position.

Answers

8708 [C] (M04) AMT-P	8709 [A] (M04) AMT-P	8710 [C] (M04) AMT-P	8711 [C] (M04) AMT-P
8712 [B] (M04) AMT-P	8713 [B] (M04) AMT-P	8714 [B] (M04) AMT-P	8715 [A] (M04) AMT-G

8716. One of the best ways to increase engine power and control detonation and preignition is to

A—enrich the fuel/air mixture.
B—use water injection.
C—lean the fuel/air mixture.

A water-injection system (ADI system) is used to allow a reciprocating engine to operate with maximum power without detonation or preignition.

8717. An excessively lean fuel/air mixture may cause

A—an increase in cylinder head temperature.
B—high oil pressure.
C—backfiring through the exhaust.

An excessively lean fuel-air mixture ratio burns slowly. It is still burning when the gases are forced out past the exhaust valve.

This slow burning will cause an increase in cylinder head temperature.

8718. The density of air is very important when mixing fuel and air to obtain a correct fuel-to-air ratio. Which of the following weighs the most?

A—75 parts of dry air and 25 parts of water vapor.
B—100 parts of dry air.
C—50 parts of dry air and 50 parts of water vapor.

Water vapor weighs only about 5/8 as much as dry air. Therefore, 100 parts of dry air will weigh more than any of the other choices given with this question.

8719. A mixture ratio of 11:1 normally refers to

A—a stoichiometric mixture.
B—1 part air to 11 parts fuel.
C—1 part fuel to 11 parts air.

An air/fuel mixture ratio of 11:1 is a mixture of eleven parts of air and one part of fuel by weight.

An air-fuel ratio of 11:1 is the same as a fuel-air ratio of 0.091.

8720. The economizer system in a float-type carburetor

A—keeps the fuel/air ratio constant.
B—functions only at cruise and idle speeds.
C—increases the fuel/air ratio at high power settings.

The economizer system in a float-type carburetor increases the fuel-air mixture ratio at high power settings.

By using the rich mixture only at the higher power settings, the engine can use an economically lean mixture during all other operations.

8721. A carburetor is prevented from leaning out during quick acceleration by the

A—power enrichment system.
B—mixture control system.
C—accelerating system.

The accelerating system provides a momentarily rich mixture during quick acceleration of the engine to prevent its "leaning out."

8722. In turbine engines that utilize a pressurization and dump valve, the dump portion of the valve

A—cuts off fuel flow to the engine fuel manifold and dumps the manifold fuel into the combustor to burn just before the engine shuts down.
B—drains the engine manifold lines to prevent fuel boiling and subsequent deposits in the lines as a result of residual engine heat (at engine shutdown).
C—dumps extra fuel into the engine in order to provide for quick engine acceleration during rapid throttle advancement.

A fuel pressurization and dump valve is usually required on engines that incorporate duplex fuel nozzles to divide the flow into primary and main manifolds.

At the flow required for starting and for altitude idling, the pressurization valve is closed, and all of the fuel passes through the primary line.

At shut down, the dump valve drains the fuel manifold to prevent the fuel boiling as a result of residual engine heat. This boiling would leave solid deposits that could clog the calibrated passageways.

8723. What effect does high atmospheric humidity have on the operation of a jet engine?

A—Decreases engine pressure ratio.
B—Decreases compressor and turbine RPM.
C—Has little or no effect.

High atmospheric humidity, which affects reciprocating engine power appreciably, has little or no effect on a turbojet engine's thrust, fuel flow and RPM.

8724. What are the positions of the pressurization valve and the dump valve in a jet engine fuel system when the engine is shut down?

A—Pressurization valve closed, dump valve open.
B—Pressurization valve open, dump valve open.
C—Pressurization valve closed, dump valve closed.

When a turbojet engine is shut down, the pressurization valve is closed and the dump valve is open.

Answers

8716 [B] (M04) AMT-P	8717 [A] (M04) AMT-P	8718 [B] (M04) AMT-P	8719 [C] (M04) AMT-P
8720 [C] (M04) AMT-P	8721 [C] (M04) AMT-P	8722 [B] (M04) AMT-P	8723 [C] (M04) AMT-P
			8724 [A] (M04) AMT-P

8725. What could cause a lean mixture and high cylinder head temperature at sea level or low altitudes?

A—Mixture control valve fully closed.
B—Defective accelerating system.
C—Automatic mixture control stuck in the extended position.

If the needle valve in an automatic mixture control in a pressure-injection carburetor sticks while it is in its extended position, it can cause a lean mixture and a high cylinder head temperature.

8726. Which of the following is NOT an input parameter for a turbine engine fuel control unit?

A—Compressor inlet pressure.
B—Compressor inlet temperature.
C—Ambient humidity.

Jet engine fuel control units often monitor the engine or high-pressure compressor speed, the compressor inlet pressure and temperature, compressor discharge pressure, and burner can pressure. Ambient humidity is not a controlling factor.

8727. Detonation occurs when the fuel/air mixture

A—burns too fast.
B—ignites before the time of normal ignition.
C—is too rich.

Detonation causes high cylinder head temperatures because the fuel-air mixture burns too fast.

In detonation, the fuel-air mixture actually explodes rather than burning evenly as it should.

8728. What corrective action should be taken when a carburetor is found to be leaking fuel from the discharge nozzle?

A—Replace the needle valve and seat.
B—Raise the float level.
C—Turn the fuel off each time the aircraft is parked.

If a float-type carburetor is leaking fuel from the discharge nozzle when the engine is not running, it is probably because the needle valve in the float bowl is leaking.

The needle valve and the seat should be replaced.

8729. A major difference between the Teledyne-Continental and RSA (Precision Airmotive or Bendix) continuous flow fuel injection systems in fuel metering is that the

A—RSA system uses fuel pressure only as a metering force.
B—Continental system utilizes airflow as a metering force.
C—Continental system uses fuel pressure only as a metering force.

The RSA continuous-flow fuel injection system uses a balance of air and fuel forces to provide the correct pressure drop across the main metering jet for all engine operating conditions except idling.

The Teledyne-Continental system does not use any air metering force, but rather it produces a fuel pressure across the metering jet that is proportional to the engine speed. This accurate pressure is produced by an engine-driven vane-type pump and a calibrated bypass orifice.

8730. The function of the altitude compensating, or aneroid valve used with the Teledyne-Continental fuel injection system on many turbocharged engines is to

A—prevent an overly rich mixture during sudden acceleration.
B—prevent detonation at high altitudes.
C—provide a means of enriching the mixture during sudden acceleration.

When the throttle of a turbocharged engine is opened suddenly, the engine accelerates faster than the turbocharger is able to increase the upper deck pressure. As a result the fuel-air mixture becomes excessively rich.

The Teledyne-Continental fuel injection system solves this problem by controlling the bypass orifice with an aneroid that senses upper deck, or turbocharger discharge pressure. When the engine accelerates and the pump output pressure increases, the aneroid holds the bypass orifice open and allows the excess fuel to return to the inlet side of the pump until the upper deck pressure increases. Then, the aneroid decreases the size of the orifice and allows the pressure, and thus the flow to increase.

8731. The primary purpose of the air bleed openings used with continuous flow fuel injector nozzles is to

A—provide for automatic mixture control.
B—lean out the mixture.
C—aid in proper fuel vaporization.

The injector nozzles used with a continuous-flow fuel injection system have air bleed holes that allow air to be drawn into the nozzle to aid in the vaporization of the fuel that is being injected into the intake valve port.

Answers

8725 [C] (M04) AMT-P	8726 [C] (M04) AMT-P	8727 [A] (M04) AMT-P	8728 [A] (M04) AMT-P
8729 [C] (M04) AMT-P	8730 [A] (M04) AMT-P	8731 [C] (M04) AMT-P	

8732. During what period does the fuel pump bypass valve open and remain open?

A—When the fuel pump pressure is greater than the demand of the engine.
B—When the boost pump pressure is greater than fuel pump pressure.
C—When the fuel pump output is greater than the demand of the carburetor.

The bypass valve around the vanes of a vane-type engine-driven fuel pump opens and remains open any time the boost pump pressure is greater than the fuel pump discharge pressure.

8733. Which of the following statements concerning a centrifugal-type fuel boost pump located in a fuel supply tank is NOT true?

A—Air and fuel vapors do not pass through a centrifugal-type pump.
B—Fuel can be drawn through the impeller section of the pump when it is not in operation.
C—The centrifugal-type pump is classified as a positive displacement pump.

A centrifugal pump is not a positive-displacement pump.

A positive-displacement pump is one that moves a constant volume of fluid each time it revolves. A centrifugal pump can rotate without pumping any fuel if its discharge is obstructed.

8734. Where is the engine fuel shutoff valve usually located?

A—Aft of the firewall.
B—Adjacent to the fuel pump.
C—Downstream of the engine-driven fuel pump.

According to 14 CFR §23.995, certificated aircraft must not have their fuel shutoff valves located on the engine side of the firewall.

8735. Boost pumps in a fuel system

A—operate during takeoff only.
B—are primarily used for fuel transfer.
C—provide a positive flow of fuel to the engine pump.

Boost pumps in an aircraft fuel tank are primarily used to provide a positive flow of fuel to the engine fuel pump.

Boost pumps are also used to transfer fuel, but this is not their primary function.

8736. (Refer to Figure 7.) What is the purpose of the fuel transfer ejectors?

A—To supply fuel under pressure to the engine-driven pump.
B—To assist in the transfer of fuel from the main tank to the boost pump sump.
C—To transfer fuel from the boost pump sump to the wing tank.

The fuel transfer ejectors provide a venturi effect in the main fuel tank to facilitate the flow of fuel from the tank into the boost pump sump.

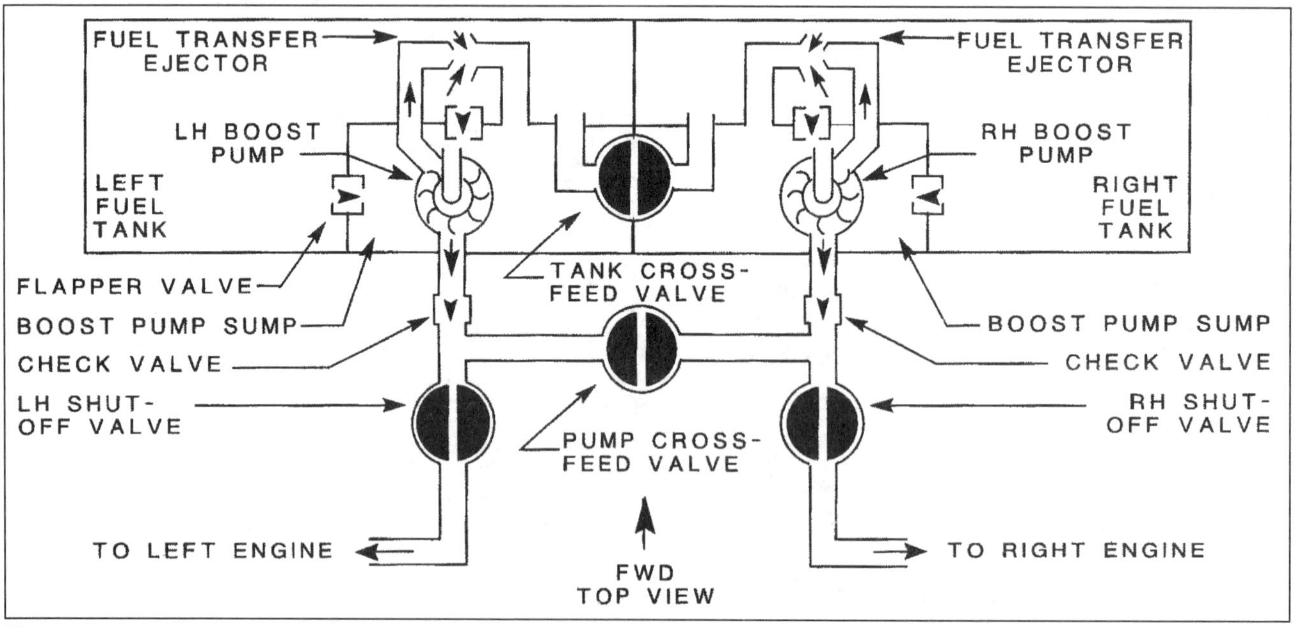

Figure 7

Answers

8732 [B] (N01) AMT-G
8733 [C] (N01) AMT-P
8734 [A] (N01) 14 CFR 23.995
8735 [C] (N01) AMT-G
8736 [B] (N01) AMT-G

8737. What is the purpose of an engine-driven fuel pump bypass valve?

A—To divert the excess fuel back to the main tank.
B—To prevent a damaged or inoperative pump from blocking the fuel flow of another pump in series with it.
C—To divert the excess fuel from the pressure side of the pump to the inlet side of the pump.

Almost all vane-type engine-driven fuel pumps have a pump bypass valve through which fuel can flow when the pump is not operating.

This bypass valve allows fuel from the booster pump to reach the engine for starting and in case the engine-driven pump fails.

8738. Most large aircraft reciprocating engines are equipped with which of the following types of engine-driven fuel pumps?

A—Rotary-vane-type fuel pump.
B—Centrifugal-type fuel pump.
C—Gear-type fuel pump.

Rotary-vane pumps are the type most generally used as fuel pumps for large reciprocating engines.

8739. When an electric primer is used, fuel pressure is built up by the

A—internal pump in the primer solenoid.
B—suction at the main discharge nozzle.
C—booster pump.

When an electric primer is used to start an aircraft engine, the fuel pressure is built up by the booster pump.

8740. The fuel pump relief valve directs excess fuel to the

A—fuel tank return line.
B—inlet side of the fuel pump.
C—inlet side of the fuel strainer.

Aircraft-engine fuel pumps have a built-in relief valve that directs the excess fuel back to the inlet side of the pump.

8741. Which type of pump is commonly used as a fuel pump on reciprocating engines?

A—Gear.
B—Impeller.
C—Vane.

The vane-type pump is the most commonly used fuel pump on reciprocating engines.

8742. The purpose of the diaphragm in most vane-type fuel pumps is to

A—maintain fuel pressure below atmospheric pressure.
B—equalize fuel pressure at all speeds.
C—compensate fuel pressures to altitude changes.

The diaphragm in a vane-type fuel pump is acted upon by the atmospheric air pressure to maintain the output pressure from the pump at a constant value above the ambient air pressure.

8743. The primary condition(s) that allow(s) microorganisms to grow in the fuel in aircraft fuel tanks is (are)

A—warm temperatures and frequent fueling.
B—the presence of water.
C—the presence of dirt or other particulate contaminants.

Microorganisms of various types can grow in an aircraft fuel tank and in storage tanks if water is present.

The microorganisms live in the water and feed on the hydrocarbon fuel.

8744. It is desirable that fuel lines have a gentle slope upward or downward and not have sharp curves or sharp rises and/or falls in order to

A—prevent vapor lock.
B—prevent stagnation or "pooling" of fuel in the fuel lines.
C—minimize the generation of static electricity by decreasing fluid friction in the lines.

Fuel lines should have sufficient size to carry the maximum required fuel flow under all operating conditions, and should have no sharp curves or sharp rises and/or falls which would trap vapors and cause vapor lock. Fuel lines should also be kept away from hot parts of the engine and exhaust system.

8745. The fuel systems of aircraft certificated in the standard classification must include which of the following?

A—An engine-driven fuel pump and at least one auxiliary pump per engine.
B—A positive means of shutting off the fuel to all engines.
C—A reserve supply of fuel, available to the engine only after selection by the flightcrew, sufficient to operate the engines at least 30 minutes at METO power.

All aircraft fuel systems must have some positive means of shutting the fuel off to all of the engines.

Answers

8737 [B] (N01) AMT-G
8738 [A] (N01) AMT-G
8739 [C] (N01) AMT-G
8740 [B] (N01) AMT-G
8741 [C] (N01) AMT-G
8742 [C] (N01) AMT-G
8743 [B] (N01) AMT-P
8744 [A] (N02) AMT-P
8745 [B] (N02) 14 CFR 23.1189

8746. Where should the main fuel strainer be located in the aircraft fuel system?

A—Downstream from the wobble pump check valve.
B—At the lowest point in the fuel system.
C—At any point in the system lower than the carburetor strainer.

The main fuel strainer in an aircraft should be located at the lowest point in the fuel system so it can collect foreign matter from the line between the tank and the engine and where it can also serve as a water trap.

8747. Where physical separation of the fuel lines from electrical wiring or conduit is impracticable, locate the fuel line

A—below the wiring and clamp the line securely to the airframe structure.
B—above the wiring and clamp the line securely to the airframe structure.
C—inboard of the wiring and clamp both securely to the airframe structure.

If fuel lines and electrical wire bundles must be routed through the same compartment, the fuel line must be below the wiring to prevent fuel from dripping into the wiring if the fuel line should leak.

The wire bundle must be securely fastened to the airframe structure, never to the fuel line.

8748. What is a characteristic of a centrifugal-type fuel boost pump?

A—It separates air and vapor from the fuel.
B—It has positive displacement.
C—It requires a relief valve.

A centrifugal-type boost pump, located in a fuel tank, separates the air and vapor from the fuel before the fuel is forced out of the tank through the fuel line.

The vapors rise to the top of the tank.

8749. The Federal Aviation Regulations require the fuel flow rate for gravity systems (main and reserve) to be

A—125 percent of the takeoff fuel consumption of the engine.
B—125 percent of the maximum, except takeoff, fuel consumption of the engine.
C—150 percent of the takeoff fuel consumption of the engine.

14 CFR §23.955(b) states that the fuel flow rate for gravity systems (main and reserve supply) must be 150% of the takeoff fuel consumption of the engine.

8750. Fuel boost pumps are operated

A—to provide a positive flow of fuel to the engine.
B—during takeoff only.
C—primarily for fuel transfer to another tank.

Fuel boost pumps are operated to provide a positive flow of fuel to the engine for starting and for high-altitude operation when there is a possibility of vapor lock.

They are also used in many installations to transfer fuel.

8751. A pilot reports that the fuel pressure fluctuates and exceeds the upper limits whenever the throttle is advanced. The most likely cause of the trouble is

A—a ruptured fuel pump relief-valve diaphragm.
B—a sticking fuel pump relief valve.
C—an air leak at the fuel pump relief-valve body.

Fluctuating fuel pressure and pressure that sometimes exceeds the upper limits for the fuel pressure can be caused by a sticky relief valve in the fuel pump.

8752. A fuel strainer or filter must be located between the

A—boost pump and tank outlet.
B—tank outlet and the fuel metering device.
C—boost pump and engine-driven fuel pump.

According to 14 CFR §23.997, there must be a fuel strainer or filter between the fuel tank outlet and the inlet of either the fuel metering device or an engine-driven, positive-displacement fuel pump, whichever is nearer the fuel tank outlet.

8753. Fuel pump relief valves designed to compensate for atmospheric pressure variations are known as

A—compensated-flow valves.
B—pressurized-relief valves.
C—balanced-type relief valves.

A balanced-type relief valve uses carburetor upper-deck pressure acting on a diaphragm in the pressure relief valve to maintain the fuel pressure at a given amount above the carburetor-inlet air pressure.

Answers

8746 [B] (N02) AMT-G	8747 [A] (N02) AC 43.13-1B	8748 [A] (N02) AMT-G	8749 [C] (N02) 14 CFR 23.9
8750 [A] (N02) AMT-SYS	8751 [B] (N02) AMT-G	8752 [B] (N02) AMT-G	8753 [C] (N02) AMT-G

8754. Fuel lines are kept away from sources of heat, and sharp bends and steep rises are avoided to reduce the possibility of

A—liquid lock.
B—vapor lock.
C—positive lock.

A vapor lock is a condition in a fuel line in which fuel vapors become trapped in a bend.

If the vapor pressure of the trapped fuel is higher than the fuel pressure, no fuel can flow to the carburetor.

Sharp bends and steep rises should be avoided, and fuel lines should not be routed in an area where the fuel can absorb heat.

8755. Fuel crossfeed systems are used in aircraft to

A—purge the fuel tanks.
B—jettison fuel in an emergency.
C—maintain aircraft stability.

The main purpose of a fuel cross-feed system in a multi-engine aircraft is to allow any engine to operate from any fuel tank.

In case one engine fails, the cross-feed system allows the fuel from all of the tanks to be used evenly. This helps maintain aircraft stability.

8756. If an engine equipped with a float-type carburetor backfires or misses when the throttle is advanced, a likely cause is that the

A—float level is too high.
B—main air bleed is clogged.
C—accelerating pump is not operating properly.

An accelerating pump is used in a float carburetor to supply a momentarily rich mixture when the throttle is suddenly opened. If the engine backfires or misses when the throttle is advanced, there is a possibility that the accelerating pump is not operating properly.

8757. A fuel pressure relief valve is required on

A—engine-driven diaphragm-type fuel pumps.
B—engine-driven vane-type fuel pumps.
C—centrifugal fuel boost pumps.

Vane-type engine-driven fuel pumps are positive-displacement pumps and require a pressure relief valve.

8758. A rotary-vane pump is best described as a

A—positive-displacement pump.
B—variable-displacement pump.
C—boost pump.

A rotary vane-type pump is a positive-displacement pump.

8759. Fuel pressure produced by the engine-driven fuel pump is adjusted by the

A—bypass valve adjusting screw.
B—relief valve adjusting screw.
C—engine-driven fuel pump adjusting screw.

The pressure produced by a vane-type engine-driven pump is adjusted by the relief valve adjusting screw.

8760. Kerosene is used as a turbine engine fuel because

A—kerosene has very high volatility which aids in ignition and lubrication.
B—kerosene has more heat energy per gallon and lubricates fuel system components.
C—kerosene does not contain any water.

Kerosene has more heat energy per gallon than gasoline, but less heat energy per pound. Kerosene also has better lubricating characteristics than aviation gasoline, and it lubricates the fuel system components.

8761. What are the principal advantages of the duplex fuel nozzle used in many turbine engines?

A—Restricts the amount of fuel flow to a level where more efficient and complete burning of the fuel is achieved.
B—Provides better atomization and uniform flow pattern.
C—Allows a wider range of fuels and filters to be used.

A duplex fuel nozzle in a turbine engine provides better atomization and a more uniform flow pattern through a wide range of engine operation than a simplex nozzle.

Answers

8754 [B] (N02) AMT-P	8755 [C] (N02) AMT-G	8756 [C] (N02) AMT-P	8757 [B] (N02) AMT-G
8758 [A] (N02) AMT-G	8759 [B] (N02) AMT-G	8760 [B] (N02) AMT-G	8761 [B] (N02) AMT-P

8762. It is necessary to control acceleration and deceleration rates in turbine engines in order to

A—prevent blowout or die-out.
B—prevent overtemperature.
C—prevent friction between turbine wheels and the case due to expansion and contraction.

The fuel control in a turbine engine controls the amount of fuel metered to the engine during acceleration and deceleration.

If too much fuel is metered to the engine when acceleration is wanted, the fuel will be discharged before the compressor can build up its speed. The fire will blow out because there is too much fuel for the amount of available air.

When the power control lever is closed, calling for a deceleration, the fuel control meters the fuel to prevent there being too little fuel for the amount of air the compressor is moving through the engine as it slows down.

8763. Which of the following turbine fuel filters has the greatest filtering action?

A—Micron.
B—Small wire mesh.
C—Stacked charcoal.

The three most common types of fuel filters used in turbine-engine fuel systems are: the micron filter, the wafer screen filter, and the plain screen mesh filter.

The micron filter has the greatest filtering action. It is capable of filtering out particles down to about 10 microns.

8764. What is the purpose of the flow divider in a turbine engine duplex fuel nozzle?

A—Allows an alternate flow of fuel if the primary flow clogs or is restricted.
B—Creates the primary and secondary fuel supplies.
C—Provides a flow path for bleed air which aids in the atomization of fuel.

A flow divider used with a turbine-engine duplex fuel nozzle creates a primary and a secondary fuel supply which are discharged through separate concentric spray tips.

This action provides the proper spray angle for all fuel flow rates.

8765. What causes the fuel divider valve to open in a turbine engine duplex fuel nozzle?

A—Fuel pressure.
B—Bleed air after the engine reaches idle RPM.
C—An electrically operated solenoid.

The flow divider valve can be a self-contained unit or it can be built into each nozzle.

Either type is a spring-loaded valve set to open at a specific fuel pressure.

8766. How often should float carburetors be overhauled?

A—At engine overhaul.
B—Annually.
C—At engine change.

There is normally no particular number of operational hours between overhauls of float carburetors, but good operating practice dictates that at the time of engine overhaul, the carburetor should be completely overhauled.

8767. What is the final authority for the details of carburetor overhaul?

A—The local FAA safety inspector.
B—The Type Certificate Data Sheets for the engine.
C—The manufacturer's recommendations.

When overhauling an aircraft carburetor, all work must be done according to the FAA-approved overhaul manual produced by the carburetor manufacturer.

8768. Excessively rich or lean idle mixtures result in

A—too rapid completion of combustion.
B—incomplete combustion.
C—incomplete cylinder scavenging.

It is important that the correct amount of fuel be metered into the air during the idling of a reciprocating engine. If the mixture is either too rich (too much fuel) or too lean (too little fuel), the mixture will not burn efficiently. There will be incomplete combustion.

8769. Which statement is true regarding proper throttle rigging of an airplane?

A—The throttle stop on the carburetor must be contacted before the stop in the cockpit.
B—The stop in the cockpit must be contacted before the stop on the carburetor.
C—The throttle control is properly adjusted when neither stop makes contact.

The throttle and mixture control must operate freely throughout their entire range of travel. The stops on the carburetor should be reached before the stops on the cockpit control. Spring-back in the control system is your assurance that the carburetor control is fully actuated.

Answers

8762 [A] (N02) AMT-P	8763 [A] (N02) AMT-P	8764 [B] (N02) AMT-P	8765 [A] (N02) AMT-P
8766 [A] (N02) AMT-P	8767 [C] (N02) AMT-P	8768 [B] (N02) AMT-P	8769 [A] (N02) AMT-P

8770. What precaution should be taken when putting thread lubricant on a tapered pipe plug in a carburetor float bowl?

A—Put the thread lubricant only on the first thread.
B—Do not use thread lubricant on any carburetor fitting.
C—Engage the first thread of the plug, then put a small amount of lubricant on the second thread and screw the plug in.

When installing tapered plugs into carburetor castings, insert the plug into its hole for one thread, then apply a small amount of thread lubricant to the second thread of the plug. Screwing the plug into the hole squeezes the lubricant between the threads and prevents galling.

8771. Maximum power is normally considered to be developed in a reciprocating engine with a fuel/air mixture ratio of approximately

A—8:1.
B—12:1.
C—15:1.

Maximum power in a reciprocating engine is normally considered to be produced with a fuel-air mixture ratio of approximately 0.083. This is the same as an air-fuel ratio of 12:1.

8772. A method commonly used to prevent carburetor icing is to

A—preheat the intake air.
B—mix alcohol with the fuel.
C—electrically heat the venturi and throttle valve.

The most commonly used method of preventing carburetor ice is to preheat the intake air.

8773. Carburetor icing is most severe at

A—air temperatures between 30 and 40°F.
B—high altitudes.
C—low engine temperatures.

Carburetor icing is most severe when the air temperature is between 30° and 40°F and the relative humidity is high.

Enough heat is required to change liquid fuel into fuel vapor to drop the temperature of the air, condense out any moisture in the air and freeze it inside the carburetor.

8774. Into what part of a reciprocating engine induction system is deicing alcohol normally injected?

A—The supercharger or impeller section.
B—The airstream ahead of the carburetor.
C—The low-pressure area ahead of the throttle valve.

Deicing alcohol is normally sprayed into the induction system of a reciprocating engine from a spray ring located at the air inlet of the carburetor.

8775. Carburetor icing on an engine equipped with a constant-speed propeller can be detected by

A—a decrease in power output with no change in manifold pressure or RPM.
B—an increase in manifold pressure with a constant RPM.
C—a decrease in manifold pressure with a constant RPM.

Carburetor ice can be detected in an engine equipped with a constant-speed propeller by a drop in manifold pressure as the ice restricts the airflow into the engine.

The constant-speed propeller will hold the engine RPM constant.

8776. What part of an aircraft in flight will begin to accumulate ice before any other?

A—Wing leading edge.
B—Propeller spinner or dome.
C—Carburetor.

The temperature drop inside the carburetor as the fuel is converted from a liquid into a vapor will cause carburetor ice to form even when there is no visible water in the air.

The other types of ice mentioned in this question are atmospheric ice in which the aircraft must be flown through visible moisture.

Carburetor ice can accumulate more readily than these other types.

8777. Carburetor icing may be eliminated by which of the following methods?

A—Alcohol spray and electrically heated induction duct.
B—Ethylene glycol spray and heated induction air.
C—Alcohol spray and heated induction air.

Carburetor ice is prevented by heating the induction air, either by routing it around the exhaust, or by picking it up in some warm portion of the engine nacelle.

Impact ice is prevented from forming on the carburetor by the use of an alcohol spray.

Answers

8770 [C] (N02) AMT-P	8771 [B] (N02) AMT-P	8772 [A] (O01) AMT-P	8773 [A] (O01) AMT-G
8774 [B] (O01) AMT-P	8775 [C] (O01) AMT-P	8776 [C] (O01) AMT-P	8777 [C] (O01) AMT-P

8778. Where would a carburetor air heater be located in a fuel injection system?

A—At the air intake entrance.
B—None is required.
C—Between the air intake and the venturi.

No carburetor air heater is required for a fuel-injected engine.

However, most engines equipped with fuel injectors have an alternate air system in which warm air from the engine compartment can be taken into the fuel injection system if ice forms on the inlet air filter.

8779. An increase in manifold pressure when carburetor heat is applied indicates

A—ice was forming in the carburetor.
B—mixture was too lean.
C—overheating of cylinder heads.

An increase in manifold pressure after carburetor heat is applied indicates that ice has been forming inside the carburetor and the heat has melted it.

8780. During full power output of an unsupercharged engine equipped with a float-type carburetor, in which of the following areas will the highest pressure exist?

A—Venturi.
B—Intake manifold.
C—Carburetor air scoop.

The venturi and the intake manifold both have a pressure lower than ambient.

Only the carburetor air scoop has a pressure higher than ambient pressure.

8781. The use of the carburetor air heater when it is not needed causes

A—a very lean mixture.
B—excessive increase in manifold pressure.
C—a decrease in power and possibly detonation.

The use of carburetor heat when it is not necessary will decrease the engine power because the engine will be operating with an excessively rich mixture.

There will also be the possibility of detonation because of the hot charge in the cylinders.

8782. As manifold pressure increases in a reciprocating engine, the

A—volume of air in the cylinder increases.
B—weight of the fuel/air charge decreases.
C—density of air in the cylinder increases.

The manifold pressure is a measure of the absolute pressure of the air inside the induction system of a reciprocating engine.

The higher the manifold pressure, the greater the density of the air being taken into the cylinders.

8783. Which of the following statements regarding volumetric efficiency of an engine is true?

A—The volumetric efficiency of an engine will remain the same regardless of the amount of throttle opening.
B—It is impossible to exceed 100 percent volumetric efficiency of any engine regardless of the type of supercharger used.
C—It is possible to exceed 100 percent volumetric efficiency of some engines by the use of superchargers of the proper type.

The volumetric efficiency of an engine is the ratio of the fuel-air charge taken into a cylinder to the charge the cylinder will hold at normal atmospheric pressure.

It is possible, by supercharging, to increase volumetric efficiency to more than 100%.

8784. Bootstrapping of a turbocharged engine is indicated by

A—an overboost condition of the engine on takeoff.
B—a transient increase in engine power.
C—a maximum increase in manifold pressure.

Bootstrapping is a transient increase in engine power that causes the turbocharger to speed up, which in turn causes the engine to produce more power. Bootstrapping is indicated by a continual drift in the manifold pressure indication.

Answers

8778 [B] (O01) AMT-P	8779 [A] (O01) AMT-P	8780 [C] (O02) AMT-P	8781 [C] (O02) AMT-P
8782 [C] (O02) AMT-P	8783 [C] (O02) AMT-P	8784 [B] (O02) AC 43.13-1B	

8785. Which of the following would be a factor in the failure of an engine to develop full power at takeoff?

A—Improper adjustment of carburetor heat valve control linkage.
B—Excessively rich setting on the idle mixture adjustment.
C—Failure of the economizer valve to remain closed at takeoff throttle setting.

Improper adjustment of the carburetor heat valve control linkage could prevent the engine from developing full takeoff power.

The density of any heated air taken into the carburetor is so low that it will not furnish enough oxygen to allow the engine to develop full power.

8786. If the turbocharger waste gate is completely closed,

A—none of the exhaust gases are directed through the turbine.
B—the turbocharger is in the OFF position.
C—all the exhaust gases are directed through the turbine.

When the waste gate of a turbosupercharger is completely closed, all of the exhaust gases flow through the turbine.

8787. Boost manifold pressure is generally considered to be any manifold pressure above

A—14.7" Hg.
B—50" Hg.
C—30" Hg.

Boost manifold pressure is normally considered to be any pressure above existing ambient pressure.

This is generally thought of as pressure above 30 inches of mercury, absolute.

8788. What is the purpose of the density controller in a turbocharger system?

A—Limits the maximum manifold pressure that can be produced at other than full throttle conditions.
B—Limits the maximum manifold pressure that can be produced by the turbocharger at full throttle.
C—Maintains constant air velocity at the carburetor inlet.

A density controller is designed to limit the manifold pressure below the turbocharger's critical altitude. It regulates bleed oil only at the full-throttle position.

8789. What is the purpose of the rate-of-change controller in a turbocharger system?

A—Limits the maximum manifold pressure that can be produced by the turbocharger at full throttle conditions.
B—Controls the rate at which the turbocharger discharge pressure will increase.
C—Controls the position of the waste gate after the aircraft has reached its critical altitude.

The rate-of-change controller controls the rate at which the turbocharger compressor discharge pressure will increase.

8790. What directly regulates the speed of a turbocharger?

A—Turbine.
B—Waste gate.
C—Throttle.

The amount of exhaust gas passing through the turbine in a turbocharger determines its speed, and the waste gate determines the amount of exhaust gas allowed to pass through the turbine.

8791. What is the purpose of a turbocharger system for a small reciprocating aircraft engine?

A—Compresses the air to hold the cabin pressure constant after the aircraft has reached its critical altitude.
B—Maintains constant air velocity in the intake manifold.
C—Compresses air to maintain manifold pressure constant from sea level to the critical altitude of the engine.

The turbocharger system used on a small aircraft compresses the air before it is taken into the induction system.

The engine is able, by using this compressed air, to automatically maintain manifold pressure from sea level to its critical altitude.

Answers

8785 [A] (O02) AMT-P	8786 [C] (O02) AMT-P	8787 [C] (O02) AMT-P	8788 [B] (O02) AMT-P
8789 [B] (O02) AMT-P	8790 [B] (O02) AMT-P	8791 [C] (O02) AMT-P	

8792. What are the three basic regulating components of a sea-level boosted turbocharger system?

1. Exhaust bypass assembly.
2. Compressor assembly.
3. Pump and bearing casing.
4. Density controller.
5. Differential pressure controller.

A—2, 3, 4.
B—1, 4, 5.
C—1, 2, 3.

The three basic regulating components in a sea-level-boosted turbocharger system are the exhaust bypass assembly, the density controller and the differential-pressure controller.

8793. The differential pressure controller in a turbocharger system

A—reduces bootstrapping during part-throttle operation.
B—positions the waste gate valve for maximum power.
C—provides a constant fuel-to-air ratio.

The differential-pressure controller reduces the undesirable condition known as bootstrapping during part-throttle operation.

Bootstrapping is an indication of unregulated power changes that result in a continual drift in manifold pressure.

8794. The purpose of a sonic venturi on a turbocharged engine is to

A—limit the amount of air that can flow from the turbocharger into the cabin for pressurization.
B—increase the amount of air that can flow from the turbocharger into the cabin for pressurization.
C—increase the velocity of the fuel/air charge.

A sonic venturi in a turbocharger installation acts as a flow limiter. When the air passing through the venturi reaches a speed of Mach 1, a shock wave forms that slows down all air passing through it. This limits the amount of air that can enter the pressurization system.

8795. What is used to drive a supercharger?

A—Exhaust gases.
B—Gear train from the crankshaft.
C—Belt drive through a pulley arrangement.

A supercharger is an air compressor that increases the amount of air an internal combustion engine can take into its induction system. There are two types of superchargers: gear driven and those driven by a turbine powered by the exhaust gases.

In modern terminology, a compressor driven by a gear train from the crankshaft is called a supercharger and one driven by the exhaust gases is called a turbosupercharger, or more simply, a turbocharger.

8796. The purpose of a bellmouth compressor inlet is to

A—provide an increased ram air effect at low airspeeds.
B—maximize the aerodynamic efficiency of the inlet.
C—provide an increased pressure drop in the inlet.

Bellmouth inlets are used on helicopters, some turboprop engines, and on engine test stands because their shape gives them a high degree of aerodynamic efficiency.

8797. What method(s) is/are used to provide clean air to the engines of helicopters and turboprop airplanes that have particle (sand and ice) separators installed?

A—Positive and negative charged areas to attract and/or repel particulates out of the airflow.
B—Air/moisture separators, and "washing" the air clean utilizing water droplets.
C—Sharp airflow directional change to take advantage of inertia and/or centrifugal force, and filters or engine inlet screens.

A centrifugal sand and ice separator removes sand and ice from the air entering the engine by forcing the inlet air to make a sharp change in its direction. Any contaminants in the air are thrown by centrifugal force into a sediment trap where they are held until they can be removed on routine maintenance.

8798. The vortex dissipators installed on some turbine-powered aircraft to prevent engine FOD utilize

A—variable inlet guide vanes (IGV) and/or variable first stage fan blades.
B—variable geometry inlet ducts.
C—a stream of engine bleed air blown toward the ground ahead of the engine.

Pod-mounted turbine engines are often so low to the ground that a vortex forms in front of the air inlet. This vortex causes sand and small stones to be drawn into the intake where they can damage the engine. Compressor bleed air is directed in a high velocity stream from a nozzle in the lower part of the engine cowling into the vortex to destroy, or dissipate it.

The compressor bleed air is shut off by a solenoid valve controlled by a switch on the landing gear when the aircraft is in the air.

Answers

8792 [B] (O02) AMT-P	8793 [A] (O02) AMT-P	8794 [A] (O02) AMT-P	8795 [B] (O02) AMT-P
8796 [B] (O02) AMT-P	8797 [C] (O02) AMT-P	8798 [C] (O02) AMT-P	

8799. Vortex dissipator systems are generally activated by

A—a landing gear switch.
B—a fuel pressure switch anytime an engine is operating.
C—an engine inlet airflow sensor.

Vortex dissipators are streams of high-velocity compressor bleed air that is directed into the vortex that forms in front of pod-mounted turbine engines.

The compressor bleed air is controlled by a solenoid valve which is actuated by a switch on the landing gear that allows the air to blow only when the aircraft is on the ground.

8800. When an engine with a subsonic divergent type inlet duct is running in place at high speed on the ground, the air pressure within the inlet is

A—negative.
B—positive.
C—ambient.

Air enters the inlet duct of a turbine engine operating on the ground at ambient pressure, but because the duct diverges, the pressure rises to a slightly higher value of positive pressure before the air enters the compressor.

8801. What indications may shift when a turbofan engine anti-icing (bleed air) system is turned on?

1. Tachometer.
2. EGT.
3. EPR.

A—1 and 2.
B—2 and 3.
C—1, 2, and 3.

When the anti-icing system is turned on in a turbofan engine there will be a slight rise in the EGT. The EPR and RPM will likely shift their indication because of the change in compression delivered to the combustor.

8802. The purpose of an engine/inlet anti-ice system is primarily to

A—remove ice from engine and/or inlet areas.
B—prevent ice formation in engine and/or inlet areas.
C—remove ice from engine and/or inlet areas and prevent ice formation in engine and/or inlet areas.

An engine inlet anti-ice system is installed on turbine-powered aircraft to prevent the formation of ice on the engine components ahead of the compressor. Any ice that forms in this area can break off and cause serious damage to the compressor.

8803. If carburetor or induction system icing is not present when carburetor heat is applied with no change in the throttle setting, the

A—mixture will become richer.
B—manifold pressure will increase.
C—engine RPM will increase.

If carburetor heat is applied when there is no icing present, the heated air, being less dense than cold air, will cause the mixture to become richer.

The use of this richer mixture will cause the engine power to decrease.

8804. When starting an engine equipped with a carburetor air heater, in what position should the heater be placed?

A—Hot.
B—Cold.
C—Neutral.

The carburetor heater should be placed in the full COLD position when the engine is being started.

Heated air is not filtered. Dirt and other contaminants can be drawn into the engine if it is operated on the ground with the carburetor air heater in the HOT position.

8805. The application of carburetor heat during engine operation will

A—decrease the weight of the fuel/air charge.
B—decrease the volume of air in the cylinder.
C—increase the density of air in the cylinder.

The application of carburetor heat will decrease the weight of the fuel-air charge taken into the engine since the heated air is less dense than cold air.

The decreased weight of the fuel-air charge will cause a decrease in the power produced by the engine.

8806. The application of carburetor heat will have which of the following effects?

A—The manifold pressure will be increased.
B—The mixture will become leaner.
C—The mixture will become richer.

The application of carburetor heat will cause the mixture ratio to become richer.

A carburetor meters fuel on the basis of the volume of air passing through its venturi, but the engine burns the fuel on the basis of the weight of the air.

Heated air is less dense (weighs less) than cold air, so the application of carburetor heat will cause the mixture to become richer.

Answers

8799 [A] (O02) AMT-P	8800 [B] (O02) AMT-P	8801 [C] (O02) AMT-P	8802 [B] (O02) AMT-P
8803 [A] (O03) AMT-P	8804 [B] (O03) AMT-P	8805 [A] (O03) AMT-P	8806 [C] (O03) AMT-P

8807. When operating an engine, the application of carburetor heat will have what effect on the fuel/air mixture?

A—Enriching the mixture because the AMC cannot make a correction for increased temperature.
B—Enriching the mixture until the AMC can make a compensation.
C—Leaning the mixture until the AMC can make a compensation.

The application of carburetor heat increases the volume of the air entering the engine (its density decreases).

Since the amount of fuel metered is a function of the volume of air entering the engine, the mixture will become richer when carburetor heat is applied.

The function of the automatic mixture control (AMC) is to compensate for this change.

8808. In addition to causing accelerated wear, dust or sand ingested by a reciprocating engine may also cause

A—silicon fouling of spark plugs.
B—sludge formation.
C—acid formation.

Sand that gets into the an engine acts as an abrasive and causes accelerated wear of the cylinder walls. Silica in the sand also forms a silicon glaze on the nose core insulators of the spark plugs. This form of contamination is an insulator at low temperature, but becomes a conductor when it is heated.

8809. In an airplane equipped with an alternate air system, if the main air duct air filter becomes blocked or clogged, the

A—system will automatically allow warm, unfiltered air to be drawn into the engine.
B—flow of air into the engine will be slowed or cut off unless alternate air is selected.
C—system will automatically allow warm, filtered alternate air to be drawn into the engine.

Some aircraft are equipped with a spring-loaded alternate air door. If the main air filter becomes clogged, the door will automatically open, allowing warm unfiltered air to be drawn into the carburetor.

8810. If a fire starts in the induction system during the engine starting procedure, what should the operator do?

A—Turn off the fuel switches to stop the fuel.
B—Continue cranking the engine.
C—Turn off all switches.

Induction-system fires in an aircraft engine are normally extinguished by keeping the engine running (or by continuing to crank it with the starter) and sucking the fire into the engine to extinguish it.

8811. On small aircraft engines, fuel vaporization may be increased by

A—cooling the air before it enters the engine.
B—circulating the fuel and air mixture through passages in the oil sump.
C—heating the fuel before it enters the carburetor.

On small aircraft reciprocating engines, fuel vaporization is improved by routing the fuel-air mixture from the carburetor through tubes in the oil sump.

Warm oil flows around the tubes and heats the fuel-air mixture.

8812. The action of a carburetor airscoop is to supply air to the carburetor, but it may also

A—cool the engine.
B—keep fuel lines cool and prevent vapor lock.
C—increase the pressure of the incoming air by ram effect.

The carburetor air scoop applies a slight positive pressure to the incoming air by the ram effect caused by the forward movement of the aircraft.

8813. A carburetor air pre-heater is not generally used on takeoff unless absolutely necessary because of the

A—loss of power and possible detonation.
B—possibility of induction system overboost.
C—inability of the engine to supply enough heat to make a significant difference.

The use of carburetor heat when not necessary can cause a loss of power and possible detonation because of the increased temperature and lower density of the incoming fuel-air charge.

Answers

8807 [B] (O03) AMT-P	8808 [A] (O03) AMT-P	8809 [A] (O03) AMT-P	8810 [B] (O03) AMT-G
8811 [B] (O03) AMT-P	8812 [C] (O03) AMT-P	8813 [A] (O03) AMT-P	

8814. The primary purpose of baffles and deflectors installed around cylinders of air-cooled aircraft engines is to

A—create a low pressure area aft of the cylinders.
B—force cooling air into close contact with all parts of the cylinders.
C—increase the volume of air used to cool the engine.

Baffles and deflectors force the cooling air into close contact with the fins of an aircraft cylinder head and cylinder barrel.

This directed flow of air aids in efficient and uniform cooling of the cylinder.

8815. What is the purpose of an augmenter used in some reciprocating engine exhaust systems?

A—To reduce exhaust back pressure.
B—To aid in cooling the engine.
C—To assist in displacing the exhaust gases.

Augmenters are venturi-shaped stainless steel tubes into which the exhaust gases from a reciprocating engine are directed.

The augmenters use the venturi effect to draw an increased airflow over the engine to augment (increase) the engine cooling.

8816. Aircraft reciprocating engine cylinder baffles and deflectors should be repaired as required to prevent loss of

A— power.
B—fin area.
C—cooling.

It is extremely important that cylinder baffles and deflectors be kept in a good state of repair. Leaking baffles or improperly oriented deflectors can cause a loss of cooling which can cause engine damage.

8817. Cracks in cooling fins that do not extend into the cylinder head may be repaired by

A—filling the extremities of crack with liquid metal.
B—removing affected area and contour filing within limits.
C—welding and then grinding or filing to original thickness.

If cooling fins are cracked, the affected area may be removed and the fins contour-filed.

This repair is permitted only if the finished repair leaves the amount of fin area on the cylinder that is specified by the engine manufacturer.

8818. Which of the following should a mechanic consult to determine the maximum amount of cylinder cooling fin that could be removed when cracks are found?

A—AC 43.13-1A.
B—Engine manufacturer's service or overhaul manual.
C—Engine structure repair manual.

Anytime there is a question about what can be done to an aircraft engine, information in the engine manufacturer's service or overhaul manual must be consulted and complied with.

8819. A bent cooling fin on an aluminum cylinder head

A—should be sawed off and filed smooth.
B—should be left alone if no crack has formed.
C—should be stop drilled or a small radius filed at the point of the bend.

A bent cooling fin on a cast-aluminum cylinder head should be left alone if it has not cracked.

Attempting to straighten the thin cast fin could cause it to crack.

8820. Where are cooling fins usually located on air-cooled engines?

A—Exhaust side of the cylinder head, inside the pistons, and connecting rods.
B—Cylinder head, cylinder walls, and inside the piston.
C—Cylinder head, cylinder barrel, and inside the piston.

Cooling fins on an air-cooled aircraft engine are located on the cylinder head, the cylinder barrel and the inside of the piston heads.

8821. How do cowl flaps aid in cooling a horizontally opposed aircraft engine?

A—Recirculates air through the engine cylinders.
B—Directs air through the engine cylinders.
C—Controls the amount of air flowing around the cylinders.

The amount of air that can flow through the cylinder fins is determined by the air-pressure differential between the top of the engine and the space below the engine. On most high-powered horizontally opposed engines, this pressure differential can be controlled with cowl flaps.

Answers

8814 [B] (P01) AMT-P
8815 [B] (P01) AMT-P
8816 [C] (P01) AMT-P
8817 [B] (P01) AMT-P
8818 [B] (P01) AMT-P
8819 [B] (P01) AMT-P
8820 [C] (P01) AMT-P
8821 [C] (P01) AMT-P

8822. The position of the cowl flaps during normal cruise flight conditions is

A—closed.
B—open.
C—one half open.

Cowl flaps should be kept fully open when the engine is run on the ground. In flight, the cowl flaps are closed as there is enough ram air flowing through the engine for adequate cooling.

8823. Generally, a small crack just started in a cylinder baffle

A—requires repair by reinforcing, such as installation of a doubler over the area.
B—requires no action unless it grows or is branched into two cracks.
C—may be stop drilled.

Small cracks in a cylinder baffle that have just started can be stop-drilled to prevent them from growing.

8824. Which of the following assists in removing heat from the metal walls and fins of an air-cooled cylinder assembly?

A—An intercooler system.
B—A baffle and cowl arrangement.
C—An engine induction system.

Baffles and a cowling arrangement are used to force cooling air between the fins of an air-cooled cylinder for better cooling.

8825. During ground operation of an engine, the cowl flaps should be in what position?

A—Fully closed.
B—Fully open.
C—Opened according to ambient conditions.

For ground operation of an aircraft reciprocating engine, the cowl flaps should be fully open.

8826. The component(s) in a turbine engine that operate(s) at the highest temperatures is/are the

A—first stage turbine nozzle guide vanes.
B—turbine disks.
C—exhaust cone.

The temperature inside a turbine engine is the highest as the gases pass through the nozzle guide vanes and first stage turbine.

8827. During an operational check of an electrically powered radial engine cowl flap system, the motor fails to operate. Which of the following is the first to be checked?

A—Flap actuator motor circuit breaker.
B—Cockpit control switch.
C—Flap actuator motor.

When any electric motor fails to operate, the first thing to check is its circuit breaker.

8828. (1) Some aircraft exhaust systems include an augmenter system to draw additional air over the engine for cooling.

(2) Augmenter systems are used to create a low pressure area at the lower rear of the aircraft engine cowling.

Regarding the above statements,

A—only No. 1 is true.
B—both No. 1 and No. 2 are true.
C—only No. 2 is true.

Statement (1) is true. The augmenter system draws additional air over the engine for cooling.

Statement (2) is also true. Augmenter tubes open into the lower rear of the aircraft engine cowling. Exhaust gases flowing through the tube create a low pressure inside the cowling that draws air from above the engine through the cooling fins.

8829. Which of the following defects would likely cause a hot spot on a reciprocating engine cylinder?

A—Too much cooling fin area broken off.
B—A cracked cylinder baffle.
C—Cowling air seal leakage.

If too much cooling-fin area is broken off of a cylinder, that particular area in the cylinder will not be properly cooled and it will develop a hot spot.

8830. What part of an air-cooled cylinder assembly has the greatest fin area per square inch?

A—Cylinder barrel.
B—Rear of the cylinder head.
C—Exhaust valve port.

The area around the exhaust valve port on the cylinder head of an air-cooled aircraft engine cylinder has the greatest fin area per square inch.

Answers
8822 [A] (P01) AMT-P
8823 [C] (P01) AMT-P
8824 [B] (P02) AMT-P
8825 [B] (P02) AMT-P
8826 [A] (P02) AMT-P
8827 [A] (P02) AMT-P
8828 [B] (P02) AMT-P
8829 [A] (P02) AMT-P
8830 [C] (P02) AMT-P

8831. Reciprocating engines used in helicopters are cooled by

A—the downdraft from the main rotor.
B—a fan mounted on the engine.
C—blast tubes on either side of the engine mount.

Helicopters cool their reciprocating engines with a fan mounted on the engine. This fan pulls air through the fins on the cylinders.

8832. The greatest portion of heat generated by combustion in a typical aircraft reciprocating engine is

A—converted into useful power.
B—carried out with the exhaust gases.
C—dissipated through the cylinder walls and heads.

Approximately 45% of the heat energy in the fuel burned in an aircraft engine is carried out the exhaust with the exhaust gases.

Another 20% is dissipated through the cylinder heads and walls, and about 5% is removed by the oil system. Only about 30% is converted into useful power.

8833. A broken cooling fin on a cylinder head

A—is cause for rejection of the head.
B—may be filed to smooth contours if damage and/or repair limits are not exceeded.
C—should be left alone.

A broken cooling fin on a cylinder head can be repaired by filing the edges to a smooth contour. This type of repair can be made only if the amount of fin area remaining after the repair is completed is within the limits allowed by the engine manufacturer.

8834. An engine becomes overheated due to excessive taxiing or improper ground runup. Prior to shutdown, operation must continue until cylinders have cooled, by running engine at

A—low RPM with oil dilution system activated.
B—idle RPM.
C—high RPM with mixture control in rich position.

If an engine is overheated from excessive taxiing or improper ground operation, it should be cooled down by operating it at idle RPM for a while before it is shut down.

8835. Cylinder head temperatures are measured by means of an indicator and a

A—resistance bulb sensing device.
B—wheatstone bridge sensing device.
C—thermocouple sensing device.

Cylinder head temperature is normally measured by a thermocouple-type system.

8836. High cylinder head temperatures are likely to result from

A—a very lean mixture at high power settings.
B—fouled spark plugs.
C—a very rich mixture at high power settings.

An excessively lean mixture will cause a high cylinder head temperature because some of the mixture will be burning when it passes out of the cylinder around the exhaust valve.

8837. The purpose of an intercooler when used with a turbocharger is to cool the

A—exhaust gases before they come in contact with the turbo drive.
B—turbocharger bearings.
C—air entering the carburetor from the turbocharger.

Intercoolers are used with large turbosuperchargers to cool the air that has been heated by compression before it enters the carburetor.

8838. Prolonged idling of an engine will usually result in

A—excessive cylinder head temperatures.
B—increased oil consumption.
C—foreign material buildup on spark plugs.

Prolonged idling of an aircraft engine will normally cause spark plugs to foul by allowing foreign material to accumulate in their firing-end cavity.

Answers

8831 [B] (P02) AMT-P	8832 [B] (P02) AMT-P	8833 [B] (P02) AMT-P	8834 [B] (P02) AMT-P
8835 [C] (P02) AMT-P	8836 [A] (P02) AMT-P	8837 [C] (P02) AMT-P	8838 [C] (P02) AMT-P

8839. The most common method and generally the best conduction of heat from the inside of a cylinder barrel to the cooling air is accomplished by

A—machining fins directly on the outside of the barrel.
B—shrinking on a jacket or muff of aluminum cooling fins around a steel cylinder sleeve.
C—machining fins directly on the outside of the barrel and shrinking on a jacket or muff of aluminum cooling fins around a steel cylinder sleeve (on different areas of the barrel).

The most widely used method of conducting heat away from the cylinder barrel of an air cooled engine is by machining fins directly on the outside of the barrel.

8840. What is the function of a blast tube as found on aircraft engines?

A—A means of cooling the engine by utilizing the propeller backwash.
B—A tube used to load a cartridge starter.
C—A device to cool an engine accessory.

A blast tube is used on an air-cooled engine to direct a stream of cooling air to some engine accessory, such as a magneto or a generator.

8841. Which statement is true regarding the air passing through the combustion section of a jet engine?

A—Most is used for engine cooling.
B—Most is used to support combustion.
C—A small percentage is frequently bled off at this point to be used for air-conditioning and/or other pneumatic powered systems.

Air passing through the combustion chambers of a turbine engine is used to support combustion as well as to cool the engine.

About 75% of the air passing through the combustion chamber is used for cooling, and only about 25% is actually involved in the combustion process.

8842. Which of the following results in a decrease in volumetric efficiency?

A—Cylinder head temperature too low.
B—Part-throttle operation.
C—Short intake pipes of large diameter.

Part-throttle operation lowers the volumetric efficiency of a reciprocating engine. It does not allow a full charge of fuel and air to be drawn into the cylinder as the piston moves down on the intake stroke.

8843. The undersides of pistons are frequently finned. The principal reason is to

A—provide sludge chambers and sediment traps.
B—provide for greater heat transfer to the engine oil.
C—support ring grooves and piston pins.

The undersides of pistons are frequently finned to provide a greater area for oil to absorb heat from the piston heads.

8844. What is the position of the cowl flaps during engine starting and warmup operations under normal conditions?

A—Full open at all times.
B—Full closed at all times.
C—Open for starting, closed for warmup.

Cowl flaps on a reciprocating engine should be open anytime the engine is operating on the ground.

Open cowl flaps provide maximum engine cooling.

8845. Increased engine heat will cause volumetric efficiency to

A—remain the same.
B—decrease.
C—increase.

Increased engine heat decreases volumetric efficiency because it lowers the density of the air entering the cylinders.

8846. Why is high nickel chromium steel used in many exhaust systems?

A—High heat conductivity and flexibility.
B—Corrosion resistance and low expansion coefficient.
C—Corrosion resistance and high heat conductivity.

High nickel-chromium steel is used in exhaust systems because of its high corrosion resistance and its low coefficient of expansion.

8847. Reciprocating engine exhaust system designs commonly used to provide for ease of installation and/or allow for expansion and contraction, may include the use of

1. spring loaded ball/flexible joints.
2. slip joints.
3. bellows.
4. flexible metal tubing.

A—1, 2, 3, and/or 4.
B—1, 2, and/or 4.
C—1, 2, and/or 3.

Answers

8839 [A] (P02) AMT-P
8843 [B] (P02) AMT-P
8840 [C] (P02) AMT-G
8844 [A] (P02) AMT-P
8841 [A] (P02) AMT-P
8845 [B] (P02) AMT-P
8842 [B] (P02) AMT-P
8846 [B] (Q01) AMT-P
8847 [C] (Q01) AMT-P

A reciprocating engine exhaust system is made in sections so that it can expand and contract with changes in temperature without cracking. The various sections of the exhaust system for a normally aspirated engine use slip joints to allow component movement. Turbocharged engines cannot tolerate the small leakage allowed by slip joints, and the sections of the exhaust system are joined with bellows and/or spring-loaded ball joints.

8848. One source commonly used for carburetor air heat is

A—turbocharger heated air.
B—alternate air heat.
C—exhaust gases.

Induction air flows over a portion of the engine exhaust system to receive heat in order to prevent the formation of carburetor ice, and to remove ice that has formed in the induction system.

8849. The hot section of a turbine engine is particularly susceptible to which of the following kind of damage?

A—Galling.
B—Pitting.
C—Cracking.

Cracking is the chief form of damage found in the hot section of a turbine engine.

The reason for the cracking is the extremes of temperature and temperature changes that exist in the hot section.

8850. What is the purpose of a slip joint in an exhaust collector ring?

A—It aids in alignment and absorbs expansion.
B—It reduces vibration and increases cooling.
C—It permits the collector ring to be installed in one piece.

The sections of an exhaust collector ring are joined together by slip joints that allow the metal to expand and contract as it is heated and cooled.

Slip joints aid in alignment and absorb expansion.

8851. Sodium-filled valves are advantageous to an aviation engine because they

A—are lighter.
B—dampen valve impact shocks.
C—dissipate heat well.

Sodium-filled valves absorb heat from inside the combustion chamber as heat is transferred into the sodium. When the sodium sloshes up into the hollow valve stem, the heat is transferred through the valve guide into the cylinder head. The heat is removed from the cylinder head by a flow of cooling air through the fins.

8852. What type of nuts are used to hold an exhaust system to the cylinders?

A—Brass or heat-resistant nuts.
B—High-temperature fiber self-locking nuts.
C—High-temperature aluminum self-locking nuts.

Brass or special high-temperature locknuts are used to hold an exhaust system to the cylinders of a reciprocating engine.

8853. Repair of exhaust system components

A—is impossible because the material cannot be identified.
B—must be accomplished by the component manufacturer.
C—is not recommended to be accomplished in the field.

It is generally recommended that exhaust stacks, mufflers, tail pipes, etc., be replaced with new or reconditioned components, rather than being repaired.

Because of the difficulty of repairing these components, they should not be repaired in the field, but should be repaired by a specially equipped and approved repair facility.

8854. On turbojet powered airplanes, thrust reversers are capable of producing between?

A—35 and 50 percent of the rated thrust in the reverse direction.
B—35 and 75 percent of the rated thrust in the reverse direction.
C—35 and 65 percent of the rated thrust in the reverse direction.

Thrust reversers provide approximately 20% of the braking force under normal runway conditions. Reversers are capable of producing between 35 and 50% of rated thrust in the reverse direction.

Answers

8848 [C] (Q01) AMT-P
8849 [C] (Q01) AMT-P
8850 [A] (Q01) AMT-P
8851 [C] (Q01) AMT-P
8852 [A] (Q01) AC 43.13-1B
8853 [C] (Q01) AC 43.13-1B
8854 [A] (Q01) AMT-P

8855. On an aircraft that utilizes an exhaust heat exchanger as a source of cabin heat, how should the exhaust system be inspected?

A—X-rayed to detect any cracks.
B—Hydrostatically tested.
C—With the heater air shroud removed.

Because of the possibility of carbon monoxide getting into the cabin through a leaking exhaust system, all of the heater muffs must be removed from the exhaust system on a maintenance inspection.

With the heater muffs removed, the entire exhaust system is carefully inspected.

8856. How should ceramic-coated exhaust components be cleaned?

A—With alkali.
B—By degreasing.
C—By mechanical means.

Ceramic-coated exhaust system components should be cleaned by degreasing only.

Ceramic-coated parts should never be cleaned by sandblasting them or by cleaning them with alkali cleaners.

8857. Which of the following indicates that a combustion chamber of a jet engine is not operating properly?

A—Clam shells stick in thrust reverse position.
B—Hot spots on the tail cone.
C—Warping of the exhaust duct liner.

Hot spots on the tail cone of a turbojet engine could indicate uneven burning within the engine.

This uneven burning could be caused by a malfunctioning combustion chamber.

8858. Select a characteristic of a good weld on exhaust stacks.

A—The weld should be built up 1/8 inch.
B—Porousness or projecting globules should show in the weld.
C—The weld should taper off smoothly into the base metal.

A weld in an exhaust system should taper smoothly into the base metal so there will be no stress concentration and no restriction to the flow of the hot exhaust gases.

8859. How do the turbines which are driven by the exhaust gases of a turbo-compound engine contribute to total engine power output?

A—By driving the crankshaft through suitable couplings.
B—By driving the supercharger, thus relieving the engine of the supercharging load.
C—By converting the latent heat energy of the exhaust gases into thrust by collecting and accelerating them.

Blowdown, or velocity, turbines used as power recovery turbines on turbocompound engines, extract energy from the exhaust gases and deliver it to the crankshaft through a fluid coupling.

8860. How should corrosion-resistant steel parts such as exhaust collectors be blast cleaned?

A—Use steel grit which has not previously been used on soft iron.
B—Use super fine granite grit.
C—Use sand which has not previously been used on iron or steel.

In cleaning corrosion-resistant steel parts with an abrasive blast, it is important that the abrasive not have any contamination of regular iron or steel in it.

Particles of iron or steel could become embedded in the corrosion-resistant steel and would eventually lead to failure of the part.

8861. Power recovery turbines used on some reciprocating engines are driven by the

A—exhaust gas pressure.
B—crankshaft.
C—velocity of the exhaust gases.

Power-recovery turbines, also known as "blowdown turbines," are velocity turbines.

PRTs are used on the Wright R-3350 Turbocompound engine to extract energy from the exhaust gases and return it to the crankshaft by means of a fluid coupling.

8862. Reciprocating engine exhaust systems that have repairs or sloppy weld beads which protrude internally are unacceptable because they cause

A—base metal fatigue.
B—localized cracks.
C—local hot spots.

Repairs with sloppy weld beads that protrude into a reciprocating-engine exhaust system are not acceptable, because they can cause local hot spots and may restrict the flow of the exhaust gases.

Answers

8855 [C] (Q02) AMT-P	8856 [B] (Q02) AMT-P	8857 [B] (Q02) AMT-P	8858 [C] (Q02) AC 43.13-1B
8859 [A] (Q02) AMT-P	8860 [C] (Q02) AMT-P	8861 [C] (Q02) AMT-P	8862 [C] (Q02) AMT-P

8863. Ball joints in reciprocating engine exhaust systems should be

A—tight enough to prevent any movement.
B—disassembled and the seals replaced every engine change.
C—loose enough to permit some movement.

When ball joints are used in the exhaust system of a reciprocating engine, it is essential that they be free to move in their sockets under all temperature conditions.

This freedom of motion prevents failure of the engine or the exhaust system.

8864. All of the following are recommended markers for reciprocating engine exhaust systems except

A—India ink.
B—lead pencil.
C—Prussian blue.

Do not mark on any exhaust system component with a lead pencil.

The graphite in the lead can make a distinct change in the molecular structure of the exhaust system material when it is heated, and it can cause the material to crack.

8865. How are combustion liner walls cooled in a gas turbine engine?

A—By secondary air flowing through the combustion chamber.
B—By the pattern of holes and louvers cut in the diffuser section.
C—By bleed air vented from the engine air inlet.

Combustion liner walls in a turbine engine are cooled by secondary air flowing through the combustion chamber.

8866. Augmenter tubes are part of which reciprocating engine system?

A—Induction.
B—Exhaust.
C—Fuel.

Augmenter tubes used on reciprocating engines are part of the exhaust system.

Exhaust gases flowing through these venturi-like tubes create a low pressure and increase the flow of cooling air across the engine.

8867. Dislodged internal muffler baffles on a small reciprocating engine may

A—obstruct the muffler outlet and cause excessive exhaust back pressure.
B—cause the engine to run excessively cool.
C—cause high fuel and oil consumption.

Internal muffler failure (baffles, diffusers, etc.) can cause partial or complete engine power loss by restricting the flow of exhaust gases.

This obstruction of the exhaust-gas flow causes an excessive amount of exhaust back pressure.

8868. What is the purpose of an exhaust outlet guard on a small reciprocating engine?

A—To prevent dislodged muffler baffles from obstructing the muffler outlet.
B—To reduce spark exit.
C—To shield adjacent components from excessive heat.

Engine power loss and excessive back pressure caused by exhaust-outlet blockage may be averted by installation of an exhaust-outlet guard, as is described in AC 43.13-1B, figures 8-21a and 8-21b.

8869. What could be a result of undetected exhaust system leaks in a reciprocating engine powered airplane?

A—Pilot/passenger incapacitation caused by carbon monoxide entering the cabin.
B—A rough-running engine with increased fuel consumption.
C—Too low exhaust back pressure resulting in the desired power settings not being attained.

Any exhaust-system leak should be regarded as a severe hazard.

Depending upon the location and type of leak, it can result in carbon monoxide (CO) poisoning of the crew and passengers, or it can cause an engine compartment fire.

Answers

8863 [C] (Q02) AMT-P	8864 [B] (Q02) AMT-P	8865 [A] (Q02) AMT-P	8866 [B] (Q02) AMT-P
8867 [A] (Q02) AC 43.13-1B	8868 [A] (Q02) AC 43.13-1B	8869 [A] (Q02) AC 43.13-1B	

8870. How may reciprocating engine exhaust system leaks be detected?

A—An exhaust trail aft of the tailpipe on the airplane exterior.
B—Fluctuating manifold pressure indication.
C—Signs of exhaust soot inside cowling and on adjacent components.

An exhaust-system leak in a reciprocating-engine exhaust system can be detected by the presence of exhaust-gas soot collecting inside the cowling and nacelle areas.

8871. Compared to normally aspirated engines, turbocharged engine exhaust systems operate at

A—similar temperatures and higher pressures.
B—higher temperatures and higher pressures.
C—similar temperatures and pressures.

The exhaust system of a turbocharged engine operates at a much higher temperature and pressure than the exhaust system of a comparable normally aspirated engine.

8872. Most exhaust system failures result from thermal fatigue cracking in the areas of stress concentration. This condition is usually caused by

A—the drastic temperature change which is encountered at altitude.
B—improper welding techniques during manufacture.
C—the high temperatures at which the exhaust system operates.

Most exhaust system failures result from thermal fatigue cracking in the areas of stress concentration.

This thermal fatigue is caused by the high temperatures at which the exhaust system operates.

8873. Thrust reversers utilizing a pneumatic actuating system usually receive operating pressure from

A—the engine bleed air system.
B—an on board hydraulic or electrical powered compressor.
C—high pressure air reservoirs.

There are three methods of actuating thrust reversers on turbine engines: pneumatic, hydraulic, and electric. The most common method is pneumatic, using engine compressor bleed air.

8874. Operating thrust reversers at low ground speeds can sometimes cause

1. sand or other foreign object ingestion.
2. hot gas re-ingestion.
3. compressor stalls.

A—1, 2, and 3.
B—1 and 2.
C—2 and 3.

Thrust reversers are used to slow the aircraft, and must be used with caution when the ground speed is low because of the danger of reingestion of hot gases and compressor stalls, and the ingestion of fine sand and other runway debris.

8875. Engines using cold stream, or both cold and hot stream reversing include

A—high bypass turbofans.
B—turbojets.
C—turbojets with afterburner.

High-bypass turbofan engines can use either cold stream or hot stream reversers, or both.

Cold-stream reversing is done by directing the output of the fan forward, and hot-stream reversing is done by directing the output of the core engine forward.

8876. The purpose of cascade vanes in a thrust reversing system is to

A—form a solid blocking door in the jet exhaust path.
B—turn the exhaust gases forward just after exiting the exhaust nozzle.
C—turn to a forward direction the fan and/or hot exhaust gases that have been blocked from exiting through the exhaust nozzle.

Cascade vanes are used in the pre-exit position of a thrust reverser to turn the escaping fan and/or hot gases that have been blocked from exiting through the exhaust nozzle to a forward direction to produce a rearward thrust.

8877. Turbojet and turbofan thrust reverser systems are generally powered by

1. fuel pressure.
2. electricity.
3. hydraulic pressure.
4. pneumatic pressure.

A—1, 3, and 4.
B—2, 3, and 4.
C—1, 2 and 3.

Turbojet and turbofan thrust reversers are generally operated by a pneumatic system, or by hydraulic or electrical power.

Answers

8870 [C] (Q02) AC 43.13-1B	8871 [B] (Q02) AMT-P	8872 [C] (Q02) AC 43.13-1B	8873 [A] (Q03) AMT-P
8874 [A] (Q03) AMT-P	8875 [A] (Q03) AMT-P	8876 [C] (Q03) AMT-P	8877 [B] (Q03) AMT-P

8878. The rearward thrust capability of an engine with the thrust reverser system deployed is

A—less than its forward capability.
B—equal to or less than its forward capability, depending on ambient conditions and system design.
C—equal to its forward capability.

Thrust reversers are normally capable of producing between 40% and 50% of the forward rated thrust.

8879. Which statement is generally true regarding thrust reverser systems?

A—It is possible to move some aircraft backward on the ground using reverse thrust.
B—Engine thrust reversers on the same aircraft usually will not operate independently of each other (must all be simultaneously).
C—Mechanical blockage system design permits a deployment position aft of the exhaust nozzle only.

It is possible to move some aircraft backward on the ground using reverse thrust in what is called a "power back" operation. Using power back to move an aircraft on the ground must be done with extreme caution, and it requires so much fuel that it is normally not economical to use.

8880. What is the proper operating sequence when using thrust reversers to slow an aircraft after landing?

A—Advance thrust levers up to takeoff position as conditions require, select thrust reverse, de-select thrust reverser, retard thrust levers to ground idle.
B—Retard thrust levers to ground idle, raise thrust reverser levers as required, and retard thrust reverser levers to ground idle.
C—Select thrust reverse, advance thrust reverser levers no higher than 75% N_1, and retard thrust reverser levers to idle at approximately normal taxi speed.

As soon as the aircraft is firmly on the ground, with the thrust levers in the ground idle position, raise the thrust reverser levers and apply as much reverse power as needed for the existing runway conditions. When the aircraft has slowed to approximately 80 knots, reduce the power to reverse idle and apply forward thrust as soon as practical.

8881. How is aircraft electrical power for propeller deicer systems transferred from the engine to the propeller hub assembly?

A—By slip rings and segment plates.
B—By slip rings and brushes.
C—By flexible electrical connectors.

Electrical power is transferred from the engine to the propeller for propeller deicing by the use of slip rings and brushes.

8882. How is anti-icing fluid ejected from the slinger ring on a propeller?

A—By pump pressure.
B—By centripetal force.
C—By centrifugal force.

Anti-icing fluid is ejected from the slinger ring on a propeller by centrifugal force as the propeller rotates.

8883. On most reciprocating multiengine aircraft, automatic propeller synchronization is accomplished through the actuation of the

A—throttle levers.
B—propeller governors.
C—propeller control levers.

Automatic propeller synchronization is nearly always accomplished by the use of propeller governors.

8884. Propeller fluid anti-icing systems generally use which of the following?

A—Ethylene glycol.
B—Isopropyl alcohol.
C—Ethyl alcohol.

The anti-icing fluid used on a propeller is normally isopropyl alcohol.

8885. What is a function of the automatic propeller synchronizing system on multiengine aircraft?

A—To control the tip speed of all propellers.
B—To control engine RPM and reduce vibration.
C—To control the power output of all engines.

Automatic propeller synchronizing systems on multi-engine aircraft hold the RPM of all of the engines the same to minimize vibration.

Answers

8878 [A] (Q03) AMT-P	8879 [A] (Q03) AMT-P	8880 [B] (Q03) AMT-P	8881 [B] (R01) AMT-P
8882 [C] (R01) AMT-P	8883 [B] (R01) AMT-P	8884 [B] (R01) AMT-P	8885 [B] (R01) AMT-P

8886. Ice formation on propellers, when an aircraft is in flight, will

A—decrease thrust and cause excessive vibration.
B—increase aircraft stall speed and increase noise.
C—decrease available engine power.

Ice formation on a propeller, when the aircraft is in flight, will alter the airfoil shape of the propeller blade.
Altering the airfoil shape decreases the thrust and causes excessive vibration.

8887. What unit in the propeller anti-icing system controls the output of the pump?

A—Pressure relief valve.
B—Rheostat.
C—Cycling timer.

Propeller anti-icing fluid is pumped to the propeller by an electric-motor-driven pump.
The output of the pump is controlled with a rheostat in the motor control circuit.

8888. Proper operation of electric deicing boots on individual propeller blades may best be determined by

A—feeling the sequence of boot heating and have an assistant observe the loadmeter indications.
B—observing the ammeter or loadmeter for current flow.
C—feeling the boots to see if they are heating.

Propeller electric deicing boots are checked for operation by observing the ammeter or loadmeter for the proper amount of current flow, and by feeling the boots to see if they are heating.
Follow the sequence of boot heating by feeling the boots with your hand to determine that each one is heating. The boots should all have a similar heat rise in the same length of time.

8889. A propeller synchrophasing system allows a pilot to reduce noise and vibration by

A—adjusting the phase angle between the propellers on an aircraft's engines.
B—adjusting the plane of rotation of all propellers.
C—setting the pitch angle of all propellers exactly the same.

Synchrophasing allows a pilot to adjust the phase angle between the propellers on the various engines to reduce the noise and vibration to a minimum.

8890. Which of the following determines oil and grease specifications for lubrication of propellers?

A—Airframe manufacturers.
B—Engine manufacturers.
C—Propeller manufacturers.

Propeller lubricant specifications are determined by the propeller manufacturer.

8891. Grease used in aircraft propellers reduces the frictional resistance of moving parts and is easily molded into any form under pressure. This statement defines

A—antifriction and plasticity characteristics of grease.
B—antifriction and chemical stability of grease.
C—viscosity and melting point of grease.

The ability of a grease to reduce friction is its anti-friction characteristic.
The ability to be easily molded into any form under pressure is a plasticity characteristic.

8892. What type of imbalance will cause a two-blade propeller to have a persistent tendency to come to rest in a horizontal position (with the blades parallel to the ground) while being checked on a propeller balancing beam?

A—Vertical.
B—Horizontal.
C—Harmonic.

If a propeller comes to rest on a balance stand with its blades in the horizontal position, one side of the propeller hub is heavier than the other.
The propeller is said to be vertically unbalanced.

8893. What is the purpose of an arbor used in balancing a propeller?

A—To support the propeller on the balance knives.
B—To level the balance stand.
C—To mark the propeller blades where weights are to be attached.

When a propeller is to be balanced, it is placed on an arbor which supports it on the balancing knife edges.

8894. If a blade of a particular metal propeller is shortened because of damage to the tip, the remaining blade(s) must be

A—reset (blade angle) to compensate for the shortened blade.
B—returned to the manufacturer for alteration.
C—reduced to conform with the shortened blade.

Answers

8886 [A] (R01) AMT-P
8887 [B] (R01) AMT-P
8888 [A] (R01) AMT-P
8889 [A] (R01) AMT-P
8890 [C] (R02) AMT-P
8891 [A] (R02) AMT-P
8892 [A] (R03) AMT-P
8893 [A] (R03) AMT-P
8894 [C] (R03) AC 43.13-1B

If one blade of a metal propeller is shortened or its shape is changed to dress out damage to the tip, the other blades must be made to conform to the shortened blade.

8895. The application of more protective coating on one blade than the other when refinishing a wood propeller

A—has little or no effect on operating characteristics.
B—should never be done.
C—may be necessary to achieve final balancing.

Final horizontal balance of a wood propeller may be accomplished by applying more protective finish to the light blade than there is on the heavy blade. A final balancing check should be made after allowing the finish to dry for at least 48 hours.

8896. Apparent engine roughness is often a result of propeller unbalance. The effect of an unbalanced propeller will usually be

A—approximately the same at all speeds.
B—greater at low RPM.
C—greater at high RPM.

The centrifugal force caused by the rotation of a propeller will cause any unbalanced condition to be felt more at high engine RPM than at low RPM.

8897. Which of the following is used to correct horizontal unbalance of a wood propeller?

A—Brass screws.
B—Shellac.
C—Solder.

If a wood propeller is found to be out of balance horizontally, the condition may be corrected by adding a small amount of solder to the metal tipping on the light blade.

8898. Propeller aerodynamic (thrust) imbalance can be largely eliminated by

A—correct blade contouring and angle setting.
B—static balancing.
C—keeping the propeller blades within the same plane of rotation.

Aerodynamic imbalance of a propeller results when the thrust of the blades is unequal. This type of imbalance can be largely eliminated by the use of the proper blade contour and blade angle setting.

8899. A powerplant using a hydraulically controlled constant-speed propeller is operating within the propeller's constant-speed range at a fixed throttle setting. If the tension of the propeller governor control spring (speeder spring) is reduced by movement of the cockpit propeller control, the propeller blade angle will

A—increase, engine manifold pressure will increase, and engine RPM will decrease.
B—decrease, engine manifold pressure will increase, and engine RPM will decrease.
C—decrease, engine manifold pressure will decrease, and engine RPM will increase.

If an engine is operated in the constant-speed range and the compression of the propeller governor speeder spring is reduced, the centrifugal force on the flyweights will produce an overspeed condition which will cause the propeller blade angle to increase.

This increase in blade angle will cause the engine RPM to decrease. If the throttle position has not been changed, the manifold pressure will increase.

8900. Why is the pulley stop screw on a propeller governor adjustable?

A—To limit the maximum engine speed during takeoff.
B—To maintain the proper blade angle for cruising.
C—To limit the maximum propeller pitch for takeoff.

The pulley stop screw on a manually adjustable constant-speed propeller governor adjusts the maximum RPM the governor allows the engine to produce.

This screw actually limits the maximum amount the speeder spring is allowed to compress.

8901. During engine operation at speeds lower than those for which the constant speed propeller control can govern in the INCREASE RPM position, the propeller will

A—remain in the full HIGH PITCH position.
B—maintain engine RPM in the normal manner until the HIGH PITCH stop is reached.
C—remain in the full LOW PITCH position.

During engine operations at speeds lower than those which the constant-speed governor can control, the propeller acts as though it were a fixed-pitch propeller in the full LOW PITCH position.

Full low pitch produces the highest RPM.

Answers

8895 [C] (R03) AC 43.13-1B	8896 [C] (R03) AMT-P	8897 [C] (R03) AMT-P	8898 [A] (R03) AMT-P
8899 [A] (R04) AMT-P	8900 [A] (R04) AMT-P	8901 [C] (R04) AMT-P	

8902. When engine power is increased, the constant-speed propeller tries to function so that it will

A—maintain the RPM, decrease the blade angle, and maintain a low angle of attack.
B—increase the RPM, decrease the blade angle, and maintain a low angle of attack.
C—maintain the RPM, increase the blade angle, and maintain a low angle of attack.

When engine power is increased in an engine equipped with a constant-speed propeller, the governor causes the blade angle to increase to hold the RPM constant.

The increase in power will possibly increase the airspeed. This will hold the propeller angle of attack low.

8903. The propeller governor controls the

A—oil to and from the pitch changing mechanism.
B—spring tension on the boost pump speeder spring.
C—linkage and counterweights from moving in and out.

A propeller governor controls the oil flowing to and from the pitch-changing mechanism of a hydraulically actuated propeller.

8904. During the on-speed condition of a propeller, the

A—centrifugal force acting on the governor flyweights is greater than the tension of the speeder spring.
B—tension on the speeder spring is less than the centrifugal force acting on the governor flyweights.
C—centrifugal force of the governor flyweights is equal to the speeder spring force.

During the on-speed condition of a propeller governor, the centrifugal force on the governor flyweights is exactly balanced by the force produced by the speeder spring.

8905. What actuates the pilot valve in the governor of a constant-speed propeller?

A—Engine oil pressure.
B—Governor flyweights.
C—Governor pump oil pressure.

The pilot valve in a propeller governor is actuated by a balance of forces between those produced by centrifugal force acting on the flyweights and the compression of the speeder spring.

8906. What action takes place when the cockpit control lever for a hydromatic, constant-speed propeller is actuated?

A—Compression of the speeder spring is changed.
B—The governor booster pump pressure is varied.
C—The governor bypass valve is positioned to direct oil pressure to the propeller dome.

When the cockpit control for a Hydromatic constant-speed propeller is actuated, the compression of the speeder spring is changed.

To increase the RPM, the compression is increased. This increased spring force requires a higher RPM to produce enough centrifugal force on the flyweights to overcome it.

8907. What will happen to the propeller blade angle and the engine RPM if the tension on the propeller governor control spring (speeder spring) is increased?

A—Blade angle will decrease and RPM will decrease.
B—Blade angle will increase and RPM will decrease.
C—Blade angle will decrease and RPM will increase.

When the tension (actually it is the compression) on the propeller governor control speeder spring is increased, the governor gives the indication of an underspeed condition.

The propeller blade angle will decrease, allowing the RPM to increase.

8908. How is the speed of a constant-speed propeller changed in flight?

A—By varying the output of the governor booster pump.
B—By advancing or retarding the throttle.
C—By changing the load tension against the flyweights in the governor.

The speed of a Hydromatic constant-speed propeller is changed by controlling the compression (load tension) of the speeder spring inside the governor.

The compression of the speeder spring opposes the centrifugal force on the flyweights in the governor.

It is the balance between the action of the flyweights and the compression of the speeder spring that determines the speed of the propeller.

8909. When the centrifugal force acting on the propeller governor flyweights overcomes the tension on the speeder spring, a propeller is in what speed condition?

A—On speed.
B—Underspeed.
C—Overspeed.

Answers

8902 [C] (R04) AMT-P	8903 [A] (R04) AMT-P	8904 [C] (R04) AMT-P	8905 [B] (R04) AMT-P
8906 [A] (R04) AMT-P	8907 [C] (R04) AMT-P	8908 [C] (R04) AMT-P	8909 [C] (R04) AMT-P

When the centrifugal force acting on the propeller governor flyweights overcomes the force of the speeder spring, the propeller is in an overspeed condition.

The governor will cause the propeller blade angle to increase to slow the propeller.

8910. What operational force causes the greatest stress on a propeller?

A— Aerodynamic twisting force.
B— Centrifugal force.
C— Thrust bending force.

The greatest stress on a propeller caused by an operational force is centrifugal force caused by the rotation of the propeller.

8911. What operational force tends to increase propeller blade angle?

A— Centrifugal twisting force.
B— Aerodynamic twisting force.
C— Thrust bending force.

Aerodynamic twisting force (ATF) tends to rotate a propeller blade to its high-pitch blade angle.

This is opposite to the effect of centrifugal twisting force (CTF).

8912. How is a propeller controlled in a large aircraft with a turboprop installation?

A— Independently of the engine.
B— By varying the engine RPM except for feathering and reversing.
C— By the engine power lever.

The turboprop fuel control and propeller governor are interconnected and operate in coordination with each other.

The engine power lever directs a signal from the cockpit to the fuel control for a specific amount of power from the engine.

The fuel control and the propeller governor together, establish the correct combination of RPM, fuel flow and propeller blade angle to create sufficient propeller thrust to provide the desired power.

8913. How does the aerodynamic twisting force affect operating propeller blades?

A— It tends to turn the blades to a high blade angle.
B— It tends to bend the blades forward.
C— It tends to turn the blades to a low blade angle.

Aerodynamic twisting force tends to turn the blades of a propeller to their high-pitch blade angle.

8914. Which of the following best describes the blade movement of a propeller that is in the high RPM position when reversing action is begun?

A— Low pitch directly to reverse pitch.
B— Low pitch through high pitch to reverse pitch.
C— Low pitch through feather position to reverse pitch.

A reversing propeller operating in the high RPM (low pitch) condition when the reversing action is begun, will go from low pitch directly to reverse pitch.

Before this can happen, the low pitch stop levers must be released to allow the propeller pitch to flatten out below its normal low pitch position.

8915. Propellers exposed to salt spray should be flushed with

A— Stoddard solvent.
B— fresh water.
C— soapy water.

Propellers that have been exposed to salt water should be flushed with fresh water until all traces of salt have been removed.

This should be accomplished as soon as possible after the salt water has splashed on the propeller.

After flushing, all parts should be thoroughly dried and all of the metal parts coated with clean engine oil or with a suitable equivalent.

8916. How can a steel propeller hub be tested for cracks?

A— By anodizing.
B— By magnetic particle inspection.
C— By etching.

A steel propeller hub can be inspected for cracks by using a magnetic particle inspection method.

8917. Which of the following functions requires the use of a propeller blade station?

A— Measuring blade angle.
B— Indexing blades.
C— Propeller balancing.

The only one of these choices that requires the use of the propeller blade stations is that of measuring the blade angle.

Blade angle is always specified at a particular blade station.

Answers

8910 [B] (R04) AMT-P	8911 [B] (R04) AMT-P	8912 [C] (R05) AMT-P	8913 [A] (R05) AMT-P
8914 [A] (R05) AMT-P	8915 [B] (R05) AMT-P	8916 [B] (R05) AMT-G	8917 [A] (R05) AMT-P

8918. The propeller blade angle is defined as the acute angle between the airfoil section chord line (at the blade reference station) and which of the following?

A—The plane of rotation.
B—The relative wind.
C—The axis of blade rotation during pitch change.

The blade angle of a propeller is defined as the acute angle between the chord of a particular blade section and the plane in which the propeller blades rotate (the plane of rotation).

8919. During which of the following conditions of flight will the blade pitch angle of a constant-speed propeller be the greatest?

A—Approach to landing.
B—Climb following takeoff.
C—High-speed, high-altitude cruising flight.

A constant-speed propeller will have the highest blade angle during high-speed, high-altitude cruising flight.

All of the other alternatives with this question require the propeller to have a low pitch angle.

8920. The actual distance a propeller moves forward through the air during one revolution is known as the

A—effective pitch.
B—geometric pitch.
C—relative pitch.

The effective pitch of a propeller is the distance the propeller actually moves through the air during one revolution.

8921. The pitch-changing mechanism of the hydromatic propeller is lubricated by

A—the pitch-changing oil.
B—using an approved-type grease in a grease gun at intervals prescribed by the propeller manufacturer.
C—thoroughly greasing, necessary only during propeller overhaul.

The dome of a Hydromatic propeller is filled with engine oil. This oil provides all of the lubrication needed for the pitch-changing mechanism.

8922. What is the result of moving the throttle on a reciprocating engine when the propeller is in the constant-speed range with the engine developing cruise power?

A—Opening the throttle will cause an increase in blade angle.
B—The RPM will vary directly with any movement of the throttle.
C—Movement of the throttle will not affect the blade angle.

If the propeller is in the constant-speed range and the engine is developing cruise power, when the throttle is advanced, instead of the RPM increasing, the propeller blade angle will increase to put an additional load on the engine and hold the RPM constant. The manifold pressure will increase.

8923. Propeller blade stations are measured from the

A—index mark on the blade shank.
B—hub centerline.
C—blade base.

Propeller blade stations are measured in inches from the center line of the propeller hub.

8924. The thrust produced by a rotating propeller is a result of

A—an area of low pressure behind the propeller blades.
B—an area of decreased pressure immediately in front of the propeller blades.
C—the angle of relative wind and rotational velocity of the propeller.

Thrust is produced by a rotating propeller in the same way lift is produced by a wing.

An area of decreased pressure is produced immediately in front of the propeller. Aerodynamic forces cause the propeller to move into this area of low pressure.

8925. Why is a constant-speed counterweight propeller normally placed in full HIGH PITCH position before the engine is stopped?

A—To prevent exposure and corrosion of the pitch changing mechanism.
B—To prevent hydraulic lock of the piston when the oil cools.
C—To prevent overheating of the engine during the next start.

The Hamilton Standard counterweight propeller is usually put in the high-pitch position before the engine is stopped.

Answers

8918 [A] (R05) AMT-P	8919 [C] (R05) AMT-P	8920 [A] (R05) AMT-P	8921 [A] (R05) AMT-P
8922 [A] (R05) AMT-P	8923 [B] (R05) AMT-P	8924 [B] (R05) AMT-P	8925 [A] (R05) AMT-P

This allows the cylinder to move back and cover the piston so it will not be exposed to moisture and other forces that could cause rust and corrosion.

On propellers in which this is not a problem, the engine should be shut down with the propeller in low pitch, instead of high pitch.

8926. The low pitch stop on a constant-speed propeller is usually set so that

A—the engine will turn at its rated takeoff RPM at sea level when the throttle is opened to allowable takeoff manifold pressure.
B—maximum allowable engine RPM cannot be exceeded with any combination of manifold pressure, altitude, or forward speed.
C—the limiting engine manifold pressure cannot be exceeded with any combination of throttle opening, altitude, or forward speed.

The low pitch stop on a constant-speed propeller is set so the engine will turn at its rated takeoff RPM at sea level when the throttle is opened, to get the allowable takeoff manifold pressure.

8927. The angle-of-attack of a rotating propeller blade is measured between the blade chord or face and which of the following?

A—Plane of blade rotation.
B—Full low-pitch blade angle.
C—Relative airstream.

The angle of attack of a propeller blade is the angle between the blade chord or face, and the relative airstream.

8928. The centrifugal twisting moment of an operating propeller tends to

A—increase the pitch angle.
B—reduce the pitch angle.
C—bend the blades in the direction of rotation.

The centrifugal twisting moment (CTM) acting on a propeller blade tends to rotate the blade to a low-pitch angle.

8929. Which of the following is identified as the cambered or curved side of a propeller blade, corresponding to the upper surface of a wing airfoil section?

A—Blade back.
B—Blade chord.
C—Blade face.

The curved side of a propeller blade (the side that corresponds to the upper surface of a wing airfoil section) is called the blade back.

The flat portion of a propeller blade is called the blade face.

8930. Which of the following best describes the blade movement of a full-feathering, constant-speed propeller that is in the LOW RPM position when the feathering action is begun?

A—High pitch through low pitch to feather position.
B—High pitch directly to feather position.
C—Low pitch through high pitch to feather position.

If a full-feathering propeller is in the low RPM, high pitch position, when the feathering action is begun, the propeller will go directly from high pitch to the feather position.

8931. The holding coil on a hydromatic propeller feathering button switch holds a solenoid relay closed that applies power to the propeller

A—governor.
B—dome feathering mechanism.
C—feathering pump motor.

The holding coil on a Hydromatic propeller feathering system holds the feather button depressed and the feathering pump operating until the propeller is feathered. Then, the oil pressure builds up high enough to open the oil pressure switch in the governor.

8932. What is the primary purpose of the metal tipping which covers the blade tips and extends along the leading edge of each wood propeller blade?

A—To increase the lateral strength of the blade.
B—To prevent impact damage to the tip and leading edge of the blade.
C—To increase the longitudinal strength of the blade.

Metal tipping is fastened to the tips and to most of the leading edge of wood propeller blades to protect them from impact damage caused by flying particles in the air, or during landing, taxiing and takeoff.

Answers

8926 [A] (R05) AMT-P	8927 [C] (R05) AMT-P	8928 [B] (R05) AMT-P	8929 [A] (R05) AMT-P
8930 [B] (R05) AMT-P	8931 [C] (R05) AMT-P	8932 [B] (R05) AMT-P	

8933. Blade angle is an angle formed by a line perpendicular to the crankshaft and a line formed by the

A—relative wind.
B—chord of the blade.
C—blade face.

Propeller blade angle is measured in degrees and is the acute angle between the chord of the blade and the plane of rotation.

The plane of rotation is perpendicular to the crankshaft.

8934. Propeller blade station numbers increase from

A—hub to tip.
B—tip to hub.
C—leading edge to trailing edge.

Propeller blade stations increase from the hub to the tip. The center of the propeller hub is station zero.

8935. The aerodynamic force acting on a rotating propeller blade operating at a normal pitch angle tends to

A—reduce the pitch angle.
B—increase the pitch angle.
C—bend the blades rearward in the line of flight.

The aerodynamic twisting force on a propeller blade tends to put the blade in a high pitch angle.

This force is countered by the centrifugal twisting force, which tends to put the blades in a low pitch angle.

The centrifugal twisting force is considerably greater than the aerodynamic twisting force, so the resultant force between the aerodynamic and the centrifugal twisting forces is a twisting moment toward low pitch. This moment is called CTM, or centrifugal twisting moment.

8936. Which of the following forces or combination of forces operates to move the blades of a constant-speed counterweight-type propeller to the HIGH PITCH position?

A—Engine oil pressure acting on the propeller piston-cylinder arrangement and centrifugal force acting on the counterweights.
B—Centrifugal force acting on the counterweights.
C—Prop governor oil pressure acting on the propeller piston-cylinder arrangement.

A constant-speed counterweight propeller has its blades moved toward high pitch by centrifugal force acting on the blade counterweights.

As the counterweights move toward the plane of propeller rotation, the blades move into a high pitch angle.

8937. The purpose of permanently sealing and partially filling some models of McCauley propeller hubs with dyed oil is to

A—provide an always clean separate lubrication of the internal parts.
B—dampen pressure surges and prevent too rapid changes in propeller blade angle.
C—make the location of cracks readily apparent.

Some models of McCauley propellers have the hub sealed and partially filled with engine oil that is colored with a red dye. This dye acts as leak detector. Any of the dyed oil that leaks from the hub is readily visible.

8938. Which of the following best describes the blade movement of a feathering propeller that is in the HIGH RPM position when the feathering action is begun?

A—High pitch through low pitch to feather position.
B—Low pitch through reverse pitch to feather position.
C—Low pitch through high pitch to feather position.

When a feathering propeller is in its high RPM, low pitch position and is feathered, it goes from low pitch, through high pitch to the feather position.

8939. The blade angle of a fixed-pitch propeller

A—is greatest at the tip.
B—is smallest at the tip.
C—increases in proportion to the distance each section is from the hub.

The blade angle of a fixed-pitch propeller decreases as the distance from the hub increases.

At the root, the angle is greatest. At the tip, the angle is the smallest.

8940. During operational check of an aircraft using hydromatic full-feathering propellers, the following observations are made:

The feather button, after being pushed, remains depressed until the feather cycle is complete, then opens.

When unfeathering, it is necessary to manually hold the button down until unfeathering is accomplished.

A—Both feather cycle and unfeather cycle are functioning properly.
B—Both feather and unfeather cycles indicate malfunctions.
C—The feather cycle is correct. The unfeather cycle indicates a malfunction.

Answers

8933 [B] (R05) AMT-P	8934 [A] (R05) AMT-P	8935 [B] (R05) AMT-P	8936 [B] (R05) AMT-P
8937 [C] (R05) AMT-P	8938 [C] (R05) AMT-P	8939 [B] (R05) AMT-P	8940 [A] (R05) AMT-P

When the feather button of a Hydromatic full-feathering propeller is pushed in, a holding coil holds the button in, allowing the feather pump to build up enough oil pressure to feather the propeller.

When the propeller blades have moved into the feathered position, the oil pressure continues to build up until an oil-pressure cutout switch in the governor opens the circuit to the holding coil. When this circuit is broken, the feather button pops back out.

To unfeather the propeller, the feather button is again pushed in. This time, it is held to allow the pressure to build up high enough to shift the distributor valve inside the propeller dome.

As soon as the blades move out of their feathered position, the feather button is released.

8941. Inspection of propeller blades by dye-penetrant inspection is accomplished to detect

A—cracks or other defects.
B—corrosion at the blade tip.
C—torsional stress.

Aluminum alloy propeller blades are inspected for cracks or any defects that extend to the surface by the dye-penetrant inspection method.

The area to be inspected is thoroughly cleaned and covered with a penetrating liquid which seeps into all defects that reach to the surface. After the penetrant has been on the surface for the proper length of time, it is washed off and the area covered with a developer which pulls the penetrant from the defect. The defect shows up as a brilliant mark on the developer.

8942. What controls the constant-speed range of a constant-speed propeller?

A—Engine RPM.
B—Angle of climb and descent with accompanying changes in airspeed.
C—The mechanical limits in the propeller pitch range.

The constant-speed range of a constant-speed propeller is limited by the mechanical stops within the propeller.

Control within this range of propeller operation is done by the governor. When the propeller blades are against the low pitch stops, the propeller is not being governed. It is not in the constant-speed range.

Anytime the propeller is in the constant-speed range, it is controlled by oil from the governor.

8943. For takeoff, a constant-speed propeller is normally set in the

A—HIGH PITCH, high RPM position.
B—HIGH PITCH, low RPM position.
C—LOW PITCH, high RPM position.

For takeoff, a constant-speed propeller is put into its low pitch, high RPM position.

With the propeller in this position, the engine can develop its maximum power.

8944. Where are the high and low pitch stops of a Hamilton Standard constant-speed or two-position counterweight propeller located?

A—In the hub and blade assembly.
B—In the counterweight assembly.
C—In the dome assembly.

In the Hamilton Standard counterweight propellers, the high-pitch and low-pitch stops are located inside the counterweight assemblies.

In some of the more modern counterweight propellers, the high-pitch and low-pitch stops are located inside the piston. In other propellers, they are located externally as part of the pushrod.

8945. Which of the following statements about constant-speed counterweight propellers is also true when referring to two-position counterweight propellers?

A—Blade angle changes are accomplished by the use of two forces, one hydraulic and the other centrifugal.
B—Since an infinite number of blade angle positions are possible during flight, propeller efficiency is greatly improved.
C—The pilot selects the RPM and the propeller changes pitch to maintain the selected RPM.

The basic difference between a counterweight constant-speed propeller and a counterweight two-position propeller is the fact that the constant-speed propeller is controlled by oil metered through a governor and the two-position propeller is controlled by oil flowing through an on-off valve.

Both propellers function with hydraulic pressure inside the cylinder taking the blades to low pitch. Centrifugal force moving the counterweights into the plane of propeller rotation increases the blade angle.

Answers
8941 [A] (R05) AC 43.13-1B
8942 [C] (R05) AMT-P
8943 [C] (R05) AMT-P
8944 [B] (R05) AMT-P
8945 [A] (R05) AMT-P

8946. Most engine-propeller combinations have one or more critical ranges within which continuous operation is not permitted. Critical ranges are established to avoid

A—severe propeller vibration.
B—low or negative thrust conditions.
C—inefficient propeller pitch angles.

Critical ranges of engine operation are ranges which include the resonant frequency of the propeller and engine combination.

At these RPMs, severe vibration may be encountered.

8947. Which of the following defects is cause for rejection of wood propellers?

A—Solder missing from screw heads securing metal tipping.
B—An oversize hub or bolthole, or elongated boltholes.
C—No protective coating on propeller.

An oversize hub or bolt hole or an elongated bolt hole in a wooden propeller, are definite reasons for rejecting the propeller.

8948. An aircraft's propeller system beta range

A—is used to produce zero or negative thrust.
B—is used to achieve maximum thrust during takeoff.
C—refers to the most fuel efficient pitch range to use at a given engine RPM.

The beta range of a turboprop propeller is used for ground operation, it includes start, taxi, and reverse operation. In this range the propeller can produce zero or negative thrust.

8949. The primary purpose of a cuff on a propeller is to

A—distribute anti-icing fluid.
B—strengthen the propeller.
C—increase the flow of cooling air to the engine nacelle.

The primary purpose for the cuff on a propeller is to increase the amount of cooling air forced into the engine cowling by the propeller.

8950. The purpose of a three-way propeller valve is to

A—direct oil from the engine oil system to the propeller cylinder.
B—direct oil from the engine through the governor to the propeller.
C—permit constant-speed operation of the propeller.

A three-way valve is used on a two-position hydraulically controlled propeller to direct engine oil into the propeller cylinder and to drain the oil from the propeller cylinder back into the engine sump.

8951. The primary purpose of a propeller is to

A—create lift on the fixed airfoils of an aircraft.
B—change engine horsepower to thrust.
C—provide static and dynamic stability of an aircraft in flight.

The primary purpose for a propeller on an airplane is to convert engine horsepower into thrust.

8952. A constant-speed propeller provides maximum efficiency by

A—increasing blade pitch as the aircraft speed decreases.
B—adjusting blade angle for most conditions encountered in flight.
C—increasing the lift coefficient of the blade.

A constant-speed propeller provides maximum efficiency by adjusting the blade angle to maintain a constant RPM as aerodynamic loads vary in flight.

8953. The centrifugal twisting force acting on a propeller blade is

A—greater than the aerodynamic twisting force and tends to move the blade to a higher angle.
B—less than the aerodynamic twisting force and tends to move the blade to a lower angle.
C—greater than the aerodynamic twisting force and tends to move the blade to a lower angle.

Centrifugal twisting force acts on a propeller blade to try to move the blade to a low pitch angle.

Centrifugal twisting force is opposite to the aerodynamic twisting force, and it is greater than the ATF.

8954. Geometric pitch of a propeller is defined as the

A—effective pitch minus slippage.
B—effective pitch plus slippage.
C—angle between the blade chord and the plane of rotation.

The geometric pitch of a propeller is the distance the propeller would advance in one revolution through a solid.

Geometric pitch is the sum of the effective pitch and the slip.

Answers

8946 [A] (R05) AMT-P
8947 [B] (R05) AMT-P
8948 [A] (R05) DAT
8949 [C] (R05) AMT-P
8950 [A] (R05) AMT-P
8951 [B] (R05) AMT-P
8952 [B] (R05) AMT-P
8953 [C] (R05) AMT-P
8954 [B] (R05) AMT-P

8955. Propeller blade angle is the angle between the

A—chord of the blade and the relative wind.
B—relative wind and the rotational plane of the propeller.
C—chord of the blade and the rotational plane of the propeller.

The propeller blade angle is the angle between the chord of the propeller blade and the rotational plane of the propeller which is perpendicular to the propeller shaft.

8956. What operational force causes propeller blade tips to lag in the opposite direction of rotation?

A—Thrust-bending force.
B—Aerodynamic-twisting force.
C—Torque-bending force.

Torque-bending force applied to a propeller blade causes the blade tips to tend to lag behind the blade in the direction opposite to the direction of rotation.

8957. What operational force tends to bend the propeller blades forward at the tip?

A—Torque-bending force.
B—Centrifugal-twisting force.
C—Thrust-bending force.

Thrust-bending force tends to bend the propeller blades forward at the tip.

8958. What are the rotational speed and blade pitch angle requirements of a constant-speed propeller during takeoff?

A—Low-speed and high-pitch angle.
B—High-speed and low-pitch angle.
C—High-speed and high-pitch angle.

A constant-speed propeller should be placed in its high-speed, low-pitch angle for takeoff.

8959. (1) A mechanic certificate with a powerplant rating authorizes the holder to repair deep scars, nicks, and dents on aluminum propeller blades.

(2) A mechanic certificate with a powerplant rating authorizes the holder to perform minor straightening of steel propeller blades.

Regarding the above statements,

A—only No. 1 is true.
B—both No. 1 and No. 2 are true.
C—neither No. 1 nor No. 2 is true.

Statement (1) is not true. A mechanic certificate with a powerplant rating does not authorize the holder to perform any major repair to a propeller. 14 CFR Part 43, Appendix A lists repairs to deep dents, cuts, scars, nicks, etc., and straightening of aluminum blades as major repairs.

Statement (2) is not true. A mechanic certificate with a powerplant rating does not authorize the holder to perform any major repair to a propeller. 14 CFR Part 43, Appendix A lists any repairs to, or straightening of steel blades as major repairs.

8960. (1) During takeoff, propeller thrust (pull) is greatest if the blade angle of attack is low and the engine power setting is high.

(2) With the aircraft stationary, propeller thrust is greatest if the blade angle of attack is high and the engine power setting is high.

Regarding the above statements,

A—only No. 1 is true.
B—only No. 2 is true.
C—both No. 1 and No. 2 are true.

Statement (1) is true. During takeoff, when maximum power and thrust are required, the constant-speed propeller is at a low propeller blade angle, or pitch. The low blade angle keeps the angle of attack small and efficient with respect to the relative wind. At the same time, it allows the propeller to handle a smaller mass of air per revolution. This light load allows the engine to turn at high RPM and produce maximum thrust.

Statement (2) is not true. Propeller static thrust is not greatest when the blade angle of attack is high. When the blade angle is low and the engine is turning at its maximum RPM, the thrust produced is maximum.

8961. Longitudinal (fore and aft) clearance of constant-speed propeller blades or cuffs must be at least 1/2 inch (12.7 mm) between propeller parts and stationary parts of the aircraft. This clearance is with the propeller blades

A—at takeoff pitch (maximum thrust) angle.
B—feathered or in the most critical pitch configuration.
C—at the lowest pitch angle.

Longitudinal (fore and aft) clearance of constant speed propeller blades or cuffs must be at least 1/2 inch (12.7 mm) between propeller parts and stationary parts of the aircraft. This clearance is with the propeller blades feathered or in the most critical pitch configuration.

Answers

8955 [C] (R05) AMT-P	8956 [C] (R05) AMT-P	8957 [C] (R05) AMT-P	8958 [B] (R05) AMT-P
8959 [C] (R05) 14 CFR 65.81	8960 [A] (R05) AMT-P	8961 [B] (R05) AMT-P	

8962. Constant-speed non-feathering McCauley, Hartzell, and other propellers of similar design without counterweights increase pitch angle using

A—oil pressure.
B—spring pressure.
C—centrifugal twisting moment.

Centrifugal twisting moment on all propeller blades tends to move the blades into low pitch. This force is opposed on propellers without counterweights by using engine oil pressure acting on a piston inside the hub to move the blades to high pitch.

8963. Counterweights on constant-speed propellers are generally used to aid in

A—increasing blade angle.
B—decreasing blade angle.
C—unfeathering the propellers.

Centrifugal twisting moment on all propeller blades tends to move the blades into low pitch. This force is opposed on counterweight propellers by centrifugal force acting on the counterweights. The counterweights tend to move into the plane of rotation, and this increases the blade pitch angle.

8964. When lubricating a Hartzell propeller blade with grease, to prevent damage to the blade seals, the service manual may recommend on some models to

A—pump grease into both zerk fittings for the blade simultaneously.
B—remove the seals prior to greasing and reinstall them afterwards.
C—remove one of the two zerk fittings for the blade and grease the blade through the remaining fitting.

To prevent damaging the blade seals when lubricating some Hartzell propellers, remove one of the zerk fittings, and pump grease through the other fitting until the old grease is pumped out and fresh grease comes from the hole.

8965. The primary purpose of a feathering propeller is to

A—prevent further engine damage when an engine fails in flight.
B—prevent propeller damage when an engine fails in flight.
C—eliminate the drag created by a windmilling propeller when an engine fails in flight.

The primary reason for using a feathering propeller on a multiengine aircraft is to decrease the drag that is caused by a windmilling propeller when an engine fails in flight.

8966. What normally prevents a Hartzell Compact propeller from going to feather when the engine is shut down on the ground?

A—Propeller cylinder air pressure.
B—A latch mechanism composed of springs and lock pins.
C—Accumulator provided oil pressure.

A Hartzell Compact propeller is feathered in flight by moving the propeller control fully aft. This allows the oil to drain from the propeller. When the oil drains out, the air charge in the propeller cylinder forces the blades into their full feather position. Centrifugal force caused by the rotating propeller holds the latch stop, which is composed of springs and lock pins, out of the way so the propeller can go into full feather.

When the engine is shut down on the ground, the propeller pitch control is in its forward position, and the engine is idling. The centrifugal force is insufficient to hold the latch stop back, and the air in the cylinder can only force the blade pitch to increase to the angle allowed by this stop.

8967. When running-up an engine and testing a newly installed hydromatic propeller, it is necessary to exercise the propeller by moving the governor control through its entire travel several times to

A—seat the blades fully against the low pitch stop.
B—free the dome of any entrapped air.
C—test the maximum RPM setting of the governor.

When a Hydromatic propeller has been installed on an engine, it must be exercised by operating it through its full range of travel a number of times.

Exercising the propeller removes any air trapped inside the dome and returns it with the oil into the engine oil sump.

8968. Which of the following occurs to cause front cone bottoming during propeller installation?

A—The front cone becomes bottomed in the front propeller hub cone seat before the rear propeller hub cone seat has engaged the rear cone.
B—The front cone enters the front propeller hub cone seat at an angle causing the propeller retaining nut to appear tight when it is only partially tightened.
C—The front cone contacts the ends of the shaft splines, preventing the front and rear cones from being tightened against the cone seats in the propeller hub.

Front-cone bottoming is a condition in which the front propeller cone contacts the ends of the shaft splines.

Answers

8962 [A] (R05) AMT-P	8963 [A] (R05) AMT-P	8964 [C] (R05) AMT-P	8965 [C] (R05) AMT-P
8966 [B] (R06) AMT-P	8967 [B] (R06) AMT-P	8968 [C] (R06) AMT-P	

When the front cone bottoms, the front and rear cones cannot be tightened against the cone seats. Even though sufficient torque is applied to the retaining nut, the propeller will not be tight on the cones.

Galling of the cones and cone seats will result.

8969. What is indicated when the front cone bottoms while installing a propeller?

A—Propeller-dome combination is incorrect.
B—Blade angles are incorrect.
C—Rear cone should be moved forward.

If the front cone bottoms when a propeller is being installed, a spacer should be placed behind the rear cone on the propeller shaft.

The spacer moves the rear cone forward so the front cone is able to seat against the front-cone seat in the propeller rather than against the ends of the shaft splines.

8970. How is the oil pressure delivery on a hydromatic propeller normally stopped after the blades have reached their full-feathered position?

A—Pulling out the feathering push button.
B—Electric cutout pressure switch.
C—Stop lugs in the teeth of the rotating cam.

The oil-pressure delivery to a Hydromatic propeller is automatically stopped when the blades reach their full-feather position.

When the blades reach their full-feather position, the oil pressure builds up and opens the pressure cutout switch in the governor. This stops the feather pump.

8971. The primary purpose of the front and rear cones for propellers that are installed on splined shafts is to

A—position the propeller hub on the splined shaft.
B—prevent metal-to-metal contact between the propeller and the splined shaft.
C—reduce stresses between the splines of the propeller and the splines of the shaft.

The primary purpose of the front and rear cones in a propeller installation is that of centering (positioning) the propeller hub on the splined shaft.

8972. Which of the following statements concerning the installation of a new fixed-pitch wood propeller is true?

A—If a separate metal hub is used, final track should be accomplished prior to installing the hub in the propeller.
B—NAS close-tolerance bolts should be used to install the propeller.
C—Inspect the bolts for tightness after the first flight and again after the first 25 hours of flying.

The hub bolts in a new wood propeller should be inspected for tightness after the first flight and after 25 hours of flight.

No definite time interval can be specified for checking bolt tightness, since this is affected by changes in the wood caused by the moisture content of the air where the airplane is flown and where it is stored.

8973. If propeller cones or hub cone seats show evidence of galling and wear, the most likely cause is

A—the pitch change stops were located incorrectly, causing the cone seats to act as the high pitch stop.
B—the propeller retaining nut was not tight enough during previous operation.
C—the front cone was not fully bottomed against the crankshaft splines during installation.

Galling and wear are caused by relative movement between two parts.

Galling of the propeller cones and cone seats can be caused by the propeller retaining nut not being tight enough during previous operation.

Galling can also be caused by the front cone bottoming against the crankshaft splines. The retaining nut can have enough torque applied to it during installation, but if the front cone bottoms, the propeller will not tighten on the cones.

8974. On aircraft equipped with hydraulically operated constant-speed propellers, all ignition and magneto checking is done with the propeller in which position?

A—High RPM.
B—Low RPM.
C—High pitch range.

All ignition testing is done on an engine equipped with a hydraulically operated constant-speed propeller, with the propeller pitch control in the low pitch (high RPM) position.

In this position, at speeds below the governing speed, the propeller blades are against their low pitch stops, and the propeller acts as though it were a fixed-pitch propeller.

When the propeller is operating against its low pitch stops, you can see a drop in the RPM when the magnetos are checked.

Answers

8969 [C] (R06) AMT-P
8970 [B] (R06) AMT-P
8971 [A] (R06) AMT-P
8972 [C] (R06) AC 43.13-1B
8973 [B] (R06) AMT-P
8974 [A] (R06) AMT-P

8975. Oil leakage around the rear cone of a hydromatic propeller usually indicates a defective

A—piston gasket.
B—spider-shaft oil seal.
C—dome-barrel oil seal.

Any oil that leaks out around the rear cone of a Hydromatic propeller had to leak past the spider-to-shaft oil seal.

This neoprene seal prevents oil leaking between the spider and the propeller shaft.

8976. Maximum taper contact between crankshaft and propeller hub is determined by using

A—bearing blue color transfer.
B—a micrometer.
C—a surface gauge.

The fit of the propeller hub on the crankshaft of an engine should be checked by using a liquid transfer ink such as Prussian blue.

Prussian blue is applied in a thin, even coating to the tapered end of the crankshaft. Then, with the key installed in the keyway, install the hub on the crankshaft and tighten the retaining nut to the correct installation torque.

Remove the hub and check the amount of ink that has been transferred from the crankshaft to the hub.

The ink transfer should indicate a minimum contact area of 70% of the total surface area.

8977. Propeller blade tracking is the process of determining

A—the plane of rotation of the propeller with respect to the aircraft longitudinal axis.
B—that the blade angles are within the specified tolerance of each other.
C—the positions of the tips of the propeller blades relative to each other.

Tracking a propeller is adjusting the plane of the propeller rotation so the tips of the blades follow each other in the same plane.

8978. What is the basic purpose of the three small holes (No. 60 drill) in the tipping of wood propeller blades?

A—To provide a means for inserting balancing shot when necessary.
B—To provide a means for periodically impregnating the blade with preservation materials.
C—To allow the moisture which may collect between the tipping and the wood to escape (vent the tipping).

Wooden propeller blades are vented by drilling three No. 60 holes approximately 3/16-inch deep in the tip, parallel to the longitudinal axis of the blade.

These small holes allow the moisture that condenses on the tipping between the metal and the wood to drain out or to be thrown out by centrifugal force.

8979. A fixed-pitch wooden propeller that has been properly installed and the attachment bolts properly torqued exceeds the out-of-track allowance by 1/16 inch. The excessive out-of-track condition may be corrected by

A—slightly overtightening the attachment bolts adjacent to the most forward blade.
B—discarding the propeller since out-of-track conditions cannot be corrected.
C—placing shims between the inner flange and the propeller.

An out-of-track condition of a fixed-pitch wooden propeller may be corrected by placing shims between the inner flange of the propeller hub and the propeller.

8980. Manually feathering a hydromechanical propeller means to

A—block governor oil pressure to the cylinder of the propeller.
B—port governor oil pressure to the cylinder of the propeller.
C—port governor oil pressure from the cylinder of the propeller.

Manually feathering a hydromechanical propeller, such as is found on modern multi-engine airplanes, is done by releasing the governor oil pressure and allowing the counterweights and feathering spring to feather the blades.

This is done by pulling the governor pitch control back to the limit of its travel. This opens up a port in the governor and allows the oil from the propeller to drain back into the engine.

8981. In what position is the constant-speed propeller control placed to check the magnetos?

A—Full decrease, low propeller blade pitch angle.
B—Full increase, high propeller blade pitch angle.
C—Full increase, low propeller blade pitch angle.

When checking the magnetos on an engine equipped with a constant-speed propeller, the propeller pitch control should be placed in the low propeller blade pitch angle (the full low pitch position). This is the full increase, or high RPM position.

Answers

8975 [B] (R06) AMT-P	8976 [A] (R06) AMT-P	8977 [C] (R06) AMT-P	8978 [C] (R06) AMT-P
8979 [C] (R06) AMT-P	8980 [C] (R06) AMT-P	8981 [C] (R06) AMT-P	

8982. If a flanged propeller shaft has dowel pins

A—install the propeller so that the blades are positioned for hand propping.
B—the propeller can be installed in only one position.
C—check carefully for front cone bottoming against the pins.

Many flanged propeller shafts have a short dowel that fits into a hole in the propeller hub. This dowel allows the propeller to be installed in only one position relative to the propeller shaft. This is done to prevent or minimize engine-propeller vibration.

8983. Repairs of aluminum alloy adjustable pitch propellers are not permitted to be made on which of the following propeller blade areas?

A—Shank.
B—Face.
C—Back.

No repairs are permitted to the shanks of aluminum alloy adjustable-pitch propeller blades.

8984. Which of the following methods is used to straighten a bent aluminum propeller blade that is within repairable limits?

A—Careful heating to accomplish straightening, followed by heat treatment to restore original strength.
B—Either hot or cold straightening, depending on the location and severity of damage.
C—Cold straightening only.

AC 43.13-1A, paragraph 585.c. states that bent propeller blades can be cold straightened within certain dimensional limits. This, though not stated here, must be done by an FAA-approved repair station.

AC 43.13-1B, states in paragraph 8-73.c. "Never straighten a damaged propeller. Even partial straightening of blades to permit shipment to a certificated propeller repair facility may result in hidden damage not being detected and an unairworthy propeller being returned to service."

8985. It is important that nicks in aluminum alloy propeller blades be repaired as soon as possible in order to

A—maintain equal aerodynamic characteristics between the blades.
B—eliminate stress concentration points.
C—equalize the centrifugal loads between the blades.

Nicks in an aluminum propeller blade cause a highly localized stress concentration that can be the beginning of a fatigue crack.

A fatigue crack can cause the loss of a portion of a blade.

8986. Generally, unless otherwise specified by the manufacturer, repairs of nicks, scratches, gouges, etc. on aluminum propeller blades must be made

A—parallel to the length of the blade.
B—perpendicular to the blade axis.
C—so as to return the damaged area to the original dimensions.

When dressing out nicks and scratches from an aluminum alloy blade, the damage is first removed and the edges of the repair blended into the blade surface. This is done with very fine sandpaper, making all strokes parallel to the length of the blade to prevent the possibility of leaving any horizontal scratches that could possibly cause stress concentrations.

8987. Minor surface damage located in a repairable area, but not on the leading or trailing edges of aluminum blades, may be repaired by first

A—filing with a riffle file.
B—filing with a half round or flat file.
C—rough sanding and applying a proper filler.

Damage to the face and back of a propeller can be cleaned out by using a spoon-like riffle file to remove the rough edges and form a shallow dish-shaped depression with a smooth surface.

8988. After proper removal of aluminum blade damage, the affected surface should be polished with

A—fine steel wool.
B—very fine sandpaper.
C—powdered soapstone.

When dressing out nicks and scratches from an aluminum alloy blade, the damage is first removed and the edges of the repair blended into the blade surface with very fine sandpaper, making all strokes parallel to the length of the blade to prevent the possibility of leaving any horizontal scratches that could possibly cause stress concentrations.

Answers

8982 [B] (R06) AMT-P	8983 [A] (R07) AC 43.13-1B	8984 [C] (R07) AC 43.13-1A	8985 [B] (R07) AMT-P
8986 [A] (R07) AMT-P	8987 [A] (R07) AMT-P	8988 [B] (R07) AMT-P	

8989. When preparing a propeller blade for inspection it should be cleaned with

A—mild soap and water.
B—steel wool.
C—methyl ethyl ketone.

Before inspecting a propeller, wash it with a solution of mild soap and water to remove all of the dirt, grease, and grass stains.

8990. What method would be used to inspect an aluminum propeller blade when a crack is suspected?

A—use a bright light.
B—magnetic particle.
C—dye-penetrant.

When an aluminum propeller blade is suspected of being cracked, it should be cleaned thoroughly and the suspect area inspected by the dye-penetrant method or by local etching.

8991. Removal of propeller blade tips within Type Certificate Data Sheet limits when correcting a defect is

A—a major alteration.
B—a major repair.
C—permitted under the privileges and limitations of a powerplant rating.

Removal of propeller blade tips within Type Certificate Data Sheet limits shortens the blades. Shortening propeller blades is a major repair and can be done only by the propeller manufacturer or by an FAA-certificated repair station approved for the particular propeller.

8992. Surface treatment to counter the effects of dye-penetrant inspection on a propeller is accomplished by

A—washing off with solvent.
B—wiping with alcohol.
C—rinse the blade in alodine solution.

After a part has been inspected by the dye-penetrant method, all of the penetrant must be removed from the surface. Water-soluble penetrants are flushed with a spray of water. Emulsifying penetrant is removed by applying an emulsifier and then washing it off with water.

A popular type of penetrant is not water soluble, neither is it emulsifiable. It must be removed by washing it off with a solvent that is specifically recommended for this type of penetrant.

8993. One of the advantages of inspecting an aluminum propeller utilizing dye-penetrant inspection procedure is that

A—defects just below the surface are indicated.
B—it shows whether visible lines and other marks are actually cracks rather than scratches.
C—it indicates overspeed condition.

Dye penetrant can be used to determine whether an indication is a scratch or a crack. Apply the penetrant and the developer to locate the defect. Sand away the resulting indication and reapply the developer. If the indication is actually a crack, more penetrant will be drawn from it and the indication will reappear.

8994. The primary reason for careful inspection and prompt repairing of minor surface defects such as scratches, nicks, gouges, etc. on aluminum alloy propellers is to prevent

A—corrosion.
B—unbalanced aerodynamics.
C—fatigue failure.

It is important that all scratches, nicks, and gouges in an aluminum alloy propeller blade be repaired as soon as they are discovered. This prevents stress concentrations that can cause fatigue failure of the blade under the high vibratory and centrifugal loads to which the propeller is subjected.

8995. Which of the following generally renders an aluminum alloy propeller unrepairable?

A—Any repairs that would require shortening and re-contouring of blades.
B—Any slag inclusions or cold shuts.
C—Transverse cracks of any size.

Transverse cracks (cracks across the blade) of any size are cause for rejection of an aluminum alloy propeller blade. Such cracks can easily cause the blade to break.

Answers

8989 [A] (R07) AMT-P	8990 [C] (R07) AMT-P	8991 [B] (R07) 14 CFR 43 & 65.81	8992 [A] (R07) AMT-P
8993 [B] (R07) AMT-P	8994 [C] (R07) AMT-P		8995 [C] (R07) AMT-P

8996. Cold straightening a bent aluminum propeller blade may be accomplished by

A—the holder of a mechanic certificate with a powerplant rating.
B—an appropriately rated repair station or the manufacturer.
C—a person working under the supervision of the holder of a mechanic certificate with both airframe and powerplant ratings.

Repairs to deep dents, cuts, scars, nicks, etc., and straightening of aluminum blades are considered to be propeller major repairs and can be done only by the manufacturer of the propeller or by an FAA-certificated repair station approved for the particular propeller.

8997. Frequently, an aircraft's auxiliary power unit (APU) generator

A—is identical to the engine-driven generators.
B—supplements the aircraft's engine-driven generators during peak loads.
C—has a higher load capacity than the engine-driven generators.

The power produced by an APU must be the same as that produced by the engine-driven generators, and APU generators are often identical to those driven by the engines.

8998. Fuel is normally supplied to an APU from

A—its own independent fuel supply.
B—the airplane's reserve fuel supply.
C—the airplane's main fuel supply.

The fuel used by an APU is normally taken from the airplane's main fuel supply.

8999. An APU is usually rotated during start by

A—a turbine impingement system.
B—a pneumatic starter.
C—an electric starter.

The small gas turbine engine that provides electrical power and compressed air is started by a DC electric starter.

9000. The function of an APU air inlet plenum is to

A—increase the velocity of the air before entering the compressor.
B—decrease the pressure of the air before entering the compressor.
C—stabilize the pressure of the air before it enters the compressor.

The APU air inlet plenum is a compartment, or chamber, into which the air flows before it enters the APU compressor. The plenum stabilizes the air pressure.

9001. When in operation, the speed of an APU

A—is controlled by a cockpit power lever.
B—remains at idle and automatically accelerates to rated speed when placed under load.
C—remains at or near rated speed regardless of the load condition.

The fuel control system used on an APU automatically controls the fuel for starting and then brings the engine up smoothly to its rated RPM. It automatically maintains this RPM as the air or electrical load conditions change.

9002. Generally, when maximum APU shaft output power is being used in conjunction with pneumatic power

A—pneumatic loading will be automatically modulated to maintain a safe EGT.
B—electrical loading will be automatically modulated to maintain a safe EGT.
C—temperature limits and loads must be carefully monitored by the operator to maintain a safe EGT.

APUs can normally supply all of the required electrical and pneumatic power. When the maximum shaft output power is being used, the pneumatic load is modulated, or varied, to keep the EGT within the safe range.

9003. When necessary, APU engine cooling before shutdown may be accomplished by

A—unloading the generator(s).
B—closing the bleed air valve.
C—opening the bleed air valve.

The APU control contains a time delay that closes the bleed air valve to remove the largest part of the APU load and allows the APU to cool down before it is shut down.

Answers

8996 [B] (R07) 14 CFR 43	8997 [A] (T01) TCAS	8998 [C] (T01) TCAS	8999 [C] (T01) TCAS
9000 [C] (T01) DAT	9001 [C] (T01) AGTP	9002 [A] (T01) TCAS	9003 [B] (T01) TCAS

9004. Usually, most of the load placed on an APU occurs when

A—an electrical load is placed on the generator(s).
B—the bleed air valve is opened.
C—the bleed air valve is closed.

The majority of the load on an APU is the pneumatic load, and opening the bleed air valve places this load on the APU.

9005. Fuel scheduling during APU start and under varying pneumatic bleed and electrical loads is maintained

A—manually through power control lever position.
B—automatically by the APU fuel control system.
C—automatically by an aircraft main engine fuel control unit.

The fuel control system used on an APU automatically controls the fuel for starting and then brings the engine up smoothly to its rated RPM. It automatically maintains this RPM as the air or electrical load conditions change.

9006. On APU's equipped with a free turbine and load compressor, the primary function of the load compressor is to

A—provide air for combustion and cooling in the engine gas path.
B—provide bleed air for aircraft pneumatic systems.
C—supply the turning force for operation of the APU generator(s).

APUs that have a free turbine and load compressor are actually turboshaft engines in which a free turbine drives a compressor that is totally independent of the APU gas-generator compressor. The bleed air from the load compressor supplies pneumatic power for starting the main engines and other airplane functions.

Answers
9004 [B] (T01) TCAS
9005 [B] (T01) AGTP
9006 [B] (T01) TCAS

Powerplant Oral & Practical Study Guide

PREPARATION & STUDY MATERIALS

Preparation

After you have successfully passed all the sections of the FAA knowledge test with a grade of at least 70%, you can bring the Airman Test Report to an FAA Designated Mechanic Examiner (DME) for your oral and practical tests.

The knowledge tests are used to assure the FAA that you have the required level of knowledge, and the oral and practical tests are used to determine that you have the basic skills to perform practical projects on the subjects that are covered by the knowledge tests.

When you schedule your oral and practical tests, get a copy of FAA Form 8610-2 Airman Certificate and/or Rating Application. Find out what tools you will need and what the DME charges to give you the tests.

When you take the oral and practical tests, give the examiner this completed form, along with the Airman Test Report showing that you have successfully passed all of the appropriate sections of the knowledge tests.

The actual oral and practical tests vary with the examiner, but they all contain questions and projects that are based on the same subjects covered in the knowledge tests. The time required for these tests varies, but six hours is considered to be standard for the tests.

The examiner will evaluate your responses with either a satisfactory or unsatisfactory grade and will record the question number on the back of your Form 8610-2. If, in his/her opinion, you do not have the needed skills in a subject, he or she will mark that section "Fail." If you fail any section, you must wait for at least 30 days before you retake that section of the test; or, if you get at least five hours of additional instruction on this subject by a licensed mechanic holding the rating you are testing for, you may retake it sooner. This additional instruction must be verified by a signed statement for the examiner so that he or she can retest you sooner than the 30-day waiting period.

After you satisfactorily pass all sections of the oral and practical tests, the examiner will send your completed Form 8610-2 to the FAA in Oklahoma City, Oklahoma, and issue you a temporary Mechanic Certificate that is good for 120 days. Before the expiration of this time, you should have your permanent certificate.

Study Materials

The ASA *Fast-Track Test Guides*, and Prepware for the General, Airframe, and Powerplant AMT have been prepared especially to help you get ready to take your FAA knowledge tests. Since the same material is covered in your oral and practical tests, review all of the questions and answers in the knowledge test portion of the Guides.

The questions that are included in the oral and practical test portion of this Guide are typical of those you are likely to be asked. The practical projects listed in each section are typical of those the examiner will likely use to check your level of skill. The actual questions and projects will depend upon the examiner.

Your examiner is a knowledgeable technician who can evaluate your capabilities, so don't try to "snow" him or her with words when you don't know the answer, and don't attempt any project that you are not competent to handle. It is far better to admit your lack of knowledge or skill than to blunder into a project which shows that you lack the judgment to properly evaluate your capabilities.

The ASA *Aviation Maintenance Technician Series General*, *Airframe*, and *Powerplant* textbooks, and the Advisory Circular Handbooks (AC 65-9A, AC 65-12A, and AC 65-15A) have been prepared to provide an entry-level aviation maintenance technician with basic knowledge needed for certification. When any project specifies that you use the proper reference materials, you are expected to use the manufacturer's service information or documents published by the FAA. The examiner will have on hand any reference materials you are expected to use, but it is your responsibility to know what material you need and how to use it.

By passing your knowledge test, you have proven that you have the necessary level of experience and knowledge, and the oral and practical tests are used to show that you have the needed skills.

RECIPROCATING ENGINES

Study Materials

Aviation Maintenance Technician Series
Powerplant textbook — Pages 19–86
ASA, Inc.

Airframe and Powerplant Mechanics
Powerplant Handbook AC 65-12A — Pages 1–38
Federal Aviation Administration

Typical Oral Questions

1. What is the main advantage of a horizontally opposed engine over a radial engine for powering modern aircraft?

   The horizontally opposed engine has a much smaller frontal area and is easier to streamline than a radial engine.

2. How many throws are there in the crankshaft of a six-cylinder horizontally opposed engine?

   Six.

3. What kind of connecting rod arrangement is used in a radial engine?

   A master rod connects the single throw of the crankshaft with a piston. All of the other pistons are connected to the master rod with link rods.

4. Of what material are most piston rings made?

   Gray cast iron.

5. What is the reason for using hydraulic valve lifters in an aircraft engine?

   Hydraulic valve lifters keep all of the clearance out of the valve operating mechanism. This decreases the wear of the valve train components.

6. At what speed does the camshaft turn, relative to the crankshaft speed in a horizontally opposed engine?

   The camshaft turns at one half of the crankshaft speed.

7. What kind of main bearings are used in a horizontally opposed engine?

   Steel-backed, lead-alloy bearing inserts.

8. On what stroke is the piston of a reciprocating engine when the intake valve begins to open?

   On the exhaust stroke.

9. On what stroke is the piston of a reciprocating engine when the exhaust valve begins to open?

   On the power stroke.

10. Why are both the hot and cold valve clearances given for most radial engines?

    The hot clearance is given for valve timing purposes. The timing is adjusted with the valves in cylinder number one, set with the hot clearance. When the timing is set, all of the valves are adjusted to their cold clearance.

11. What is meant by a cam-ground piston?

    A piston that is not perfectly round. Its dimension parallel with the wrist pin is several thousandths of an inch less than its dimension perpendicular to the wrist pin. When the piston reaches operating temperature, the metal in the piston pin boss expands enough that the piston becomes perfectly round.

12. Where is the piston in a reciprocating engine when the ignition spark occurs?

    About 30 degrees of crankshaft rotation before the piston reaches top center on the compression stroke.

13. What is meant by a full-floating wrist pin?

    A wrist pin that is not clamped in either the piston or the connecting rod. Full-floating wrist pins are kept from scoring the cylinder walls by soft metal plugs in their ends.

14. Why do most aircraft reciprocating engines use more than one spring on each valve?

    By using more than one spring and having the wire diameter and pitch of the springs different, valve float is minimized. The springs have different resonant frequencies, so at least one spring will always be exerting a force on the valve.

15. Would excessive valve clearance cause the valves to open early or late?

    Excessive clearance will cause the valve to open late and close early.

16. What is the purpose of valve overlap in a reciprocating engine?

    Valve overlap allows the inertia of the exhaust gases leaving the cylinder to help the fresh induction charge start flowing into the cylinder.

17. What type of piston rings are installed on the pistons of an aircraft reciprocating engine?

    Compression rings, oil control rings, and oil wiper rings.

18. Why are some exhaust valves partially filled with metallic sodium?

    The metallic sodium melts at engine operating temperature and sloshes back and forth inside the hollow valve. It picks up heat from the valve head and transfers it into the valve stem, so it can be transferred to the cylinder head through the valve guide.

19. What causes detonation in an aircraft engine?

    Excessive heat and pressure in the engine cylinder causes the fuel-air mixture to reach its critical pressure and temperature. Under these conditions, the mixture explodes rather than burns. This explosion is called detonation.

20. Why is a compression check important for determining the condition of an aircraft reciprocating engine?

    A compression check can determine the condition of the seal between the piston rings and the cylinder walls, and the seal between the intake and exhaust valves and their seats.

21. What is meant by the compression ratio of a reciprocating engine?

    The ratio of the volume of the cylinder with the piston at the bottom of its stroke to the volume with the piston at the top of its stroke.

Typical Practical Projects

1. Correctly remove a cylinder and its piston from an engine specified by the examiner.

2. Examine the rings installed on a piston for the correct tension, end gap, and side clearance.

3. Dimensionally inspect an aircraft engine cylinder for bore diameter, out-of-round, and taper.

4. Inspect the valves from an aircraft engine cylinder for stretch and for their fit in the valve guides.

5. Inspect valve springs for their specified compression strength.

6. Grind a valve seat in an engine cylinder, using the correct stones for grinding and for narrowing the seat.

7. Reface an aircraft valve to the recommended angle, and check its fit and seal in the valve seat.

8. Explain to the examiner the correct way to adjust the oil pressure in an aircraft reciprocating engine.

9. Using the correct measuring instruments, measure the diameter of journals of a crankshaft and determine whether or not they are within the tolerances allowed by the engine manufacturer.

10. Examine the bearings in a crankcase specified by the examiner, and determine their physical condition and whether or not they are within dimensional tolerance.

11. Examine a rocker arm of an aircraft engine by the magnetic particle inspection method.

12. Examine a cast aluminum or magnesium engine component by the dye penetrant inspection method.

13. Demonstrate to the examiner the correct way to start an aircraft reciprocating engine.

14. Demonstrate to the examiner the proper way to make an engine run-up check to determine the condition of the engine.

15. Identify the sludge plugs in the crankshaft of an aircraft engine, and explain their purpose.

16. Explain to the examiner the things that should be checked about engine shock mounts.

17. Perform a crankshaft runout inspection on an engine specified by the examiner.

18. Demonstrate to the examiner the correct way to find the top dead center position of the piston in the cylinder.

TURBINE ENGINES

Study Materials

Aviation Maintenance Technician Series
Powerplant textbook — Pages 337–415
ASA, Inc.

Airframe and Powerplant Mechanics
Powerplant Handbook AC 65-12A — Pages 38–69
Federal Aviation Administration

Typical Oral Questions

1. What are the two basic sections of a turbine engine?

   The hot section and the cold section.

2. What are the two basic types of compressors that are used in aircraft turbine engines?

   Axial-flow and centrifugal.

3. What is the purpose of the stators in an axial-flow compressor?

   The stators convert some velocity energy into pressure energy and change the direction of the air so it is proper for the next stage of rotors.

4. What are three types of combustors used on aircraft turbine engines?

   Can-type, annular-type, and can-annular-type.

5. How many igniters are there normally in an aircraft turbine engine?

   Two.

6. What is the main purpose of the turbine nozzle in an aircraft turbine engine?

   The turbine nozzle directs the hot gases as they leave the combustors so they will turn the turbine wheel with maximum efficiency.

7. What is meant by a free-turbine turboshaft engine?

   A turboshaft engine that has a turbine wheel, or stage of turbine wheels, that is not used to drive the compressor of the gas generator section of the engine. This free turbine drives the propeller in a turboprop engine or the transmission and rotor of a helicopter.

8. Why do some axial-flow turbine engines have more than one set of turbines and compressors?

   A two-spool turbine engine has a low-pressure and a high-pressure compressor, each driven by its own turbine. The two independent systems operate at the speed at which they are most efficient.

9. What kind of bearings are used to support the rotor shaft of an aircraft turbine engine?

   Anti-friction bearings, such as ball bearings or roller bearings.

10. What is meant by a turbofan engine?

    An axial-flow turbine engine in which the first stage of compressor blades are lengthened, so they can force air around the outside of the gas generator portion of the engine.

11. Where can you find the limits of repair allowed for the compressor blades of an aircraft turbine engine?

    In the FAA-approved service manual issued by the manufacturer of the engine.

12. Why is it important that a turbojet engine be allowed to cool before it is shut down after it has been operated at a high power setting?

    If the engine is shut down while it is hot, there is a possibility that the shroud will contract around the turbine wheel and seize the rotor.

13. What is meant by a hung start in a turbine engine?

    A start in which ignition occurs, but the engine does not accelerate to a self-sustaining speed.

14. What is meant by a hot start in a turbine engine?

    A start in which ignition occurs, but the internal temperatures go high enough that they can damage the engine.

15. What is meant by creep of the turbine blades?

    A condition of permanent elongation of the turbine blades. Creep is caused by the high temperatures and the high centrifugal loading imposed on the blades.

16. How is the compressor of a turbine engine cleaned?

    An emulsion-type cleaner is sprayed through the engine while it is being motored by the starter or operated at a low speed. The wash is followed by a clean water rinse.

    More vigorous cleaning is done by injecting a mild abrasive, such as ground-up apricot pits or walnut shells, into the engine while it is operating at low speed.

17. Where is water injected into a turbine engine for cooling purposes?

    Into the compressor inlet and into the engine diffuser case.

18. At what point in a turbine engine is the temperature the highest?

    At the inlet to the high-pressure turbine.

19. What is the function of the interconnect tubes between the cans of a turbine engine that uses can-type combustors?

    These tubes allow the flame to travel from the cans that contain the igniters to all of the other cans when the engine is being started.

20. What are two types of thrust reversers that are used on turbojet engines?

    The mechanical blockage-type and the aerodynamic blockage-type.

21. What is meant by trimming a turbojet engine?

    Adjusting the fuel control to get the correct idling and maximum-thrust RPM.

Typical Practical Projects

1. Demonstrate to the examiner the correct way to start, run up, and shut down an aircraft turbine engine.
2. Explain to the examiner the correct way to trim a turbine engine.
3. Explain to the examiner the correct way of replacing turbine blades in a turbine wheel, including the correct way of locking the blades in place.
4. Explain to the examiner the reason for having variable-angle inlet guide vanes on some axial-flow compressors.
5. Correctly remove and replace a burner can in a turbine engine specified by the examiner.
6. Demonstrate to the examiner the correct way to compute the torque that is produced by a torque wrench with an extension installed on it.
7. Check for the proper torque on a series of bolts on the case of a turbine engine, and correctly safety-wire them.
8. Correctly remove and replace a fuel nozzle in a turbine engine specified by the examiner.
9. Check the blades in a stage of an axial-flow compressor specified by the examiner for indications of damage and for security in the wheel.
10. Identify the various stations of a turbine engine and tell what pressure and temperature exist at each station during engine operation.
11. Identify the point at which bleed air is taken from a turbine engine.
12. Inspect the insulating blanket around a turbine engine for proper security and for indications of damage.

ENGINE INSPECTION

Study Materials

Aviation Maintenance Technician Series Powerplant textbook   Pages 285–297 and 520–530
ASA, Inc.

Airframe and Powerplant Mechanics Powerplant Handbook AC 65-12A   Pages 411–500
Federal Aviation Administration

Title 14 of the Code of Federal Regulations, Part 43
Federal Aviation Administration

Title 14 of the Code of Federal Regulations, Part 91
Federal Aviation Administration

Typical Oral Questions

1. What determines whether or not a 100-hour inspection is required for an aircraft?

   100-hour inspections are required only on aircraft that carry persons for hire or are used to give flight instruction for hire.

2. Where can you find a list of items that should be inspected on a 100-hour and an annual inspection of an aircraft engine?

   In Appendix D of 14 CFR Part 43.

3. In what publication can a mechanic find the operating limitations for an aircraft engine?

   In the Type Certificate Data Sheets for the engine.

4. Who is authorized to perform the powerplant inspection that is required for an annual inspection?

   An A&P mechanic who holds an Inspection Authorization.

5. What are two types of compression checks that can be performed on an aircraft reciprocating engine?

   A differential compression check and a direct compression check.

6. What items are checked in a hot section inspection of a turbine engine?

   The combustion section is checked for cracks or distortion. The turbine wheel, the turbine case, and the exhaust section are checked for cracks, indications of overheating, and any indication of warpage, erosion, or burning.

7. What is the most common type of damage that is found in the hot section of a turbine engine?

   Cracks that are caused by the high concentration of heat in the hot section.

8. Is the use of a checklist required when performing an annual or 100-hour inspection on an aircraft engine?

   Yes.

9. What inspection must be performed on an aircraft reciprocating engine if it has been operated on turbine fuel?

   The inside of the cylinders must be inspected with a borescope. The oil must be changed and the filters examined, and the engine given a careful run-up and operational check.

Typical Practical Projects

1. Perform a compression check on a reciprocating engine specified by the examiner.

2. Using the correct reference material, perform a hot-start inspection on a turbine engine.

3. Using the correct reference material, perform an overspeed inspection on a turbine engine.

4. Explain to the examiner the things that should be checked for in a hot-section inspection on a turbine engine.

5. Using the correct information, make a list of all of the Airworthiness Directives that are applicable to a particular aircraft powerplant.

6. Check the aircraft maintenance records to determine whether or not all of the applicable Airworthiness Directives have been complied with.

7. Perform a 100-hour inspection of an aircraft engine specified by the examiner and list all of the discrepancies that would cause the engine to be unairworthy.

8. Check the applicable publications and determine whether or not the engine and propeller that are installed on a particular aircraft are certificated for it.

9. Make a maintenance record entry of a 100-hour inspection that has been conducted for an aircraft engine.

10. Locate the data plate on a turbine engine and explain to the examiner the meaning of the various limitations that are listed on it.

11. Demonstrate to the examiner the correct way to connect a Jetcal analyzer to a turbine engine to measure the exhaust gas temperature and the engine RPM.

12. Perform a borescope inspection of the cylinders of a reciprocating engine.

13. Perform a borescope inspection of the internal portion of a turbine engine specified by the examiner.

ENGINE INSTRUMENT SYSTEMS

Study Materials

Aviation Maintenance Technician Series
Powerplant textbook — Pages 541–580
ASA, Inc.

Airframe and Powerplant Mechanics Powerplant Handbook AC 65-12A — Pages 483–488
Federal Aviation Administration

Typical Oral Questions

1. In what units is the tachometer for a reciprocating engine calibrated?

   In hundreds of RPM.

2. In what units is the tachometer for a turbine engine calibrated?

   In percent of the engine's rated takeoff RPM.

3. Where does the manifold pressure gage used on an aircraft reciprocating engine pick up the pressure it measures?

   From the intake manifold of the engine.

4. What are two types of pickups used for measuring the cylinder head temperature of a reciprocating engine?

   The thermocouple can be embedded in a gasket that is installed under the spark plug of the hottest running cylinder, or it can be in a bayonet that is held against the cylinder head by a spring.

5. Why do most engine oil pressure gages have a restrictor in the line between the engine and the instrument?

   The restrictor damps out pressure pulsations in the oil to keep the needle from oscillating.

6. What is used to measure the fuel flow of a horizontally opposed, fuel-injected reciprocating engine?

   A pressure gage that measures the pressure drop across injector nozzles.

7. What is indicated if the needle of a ratiometer-type oil temperature indicator pegs to the high side of the dial as soon as the aircraft master switch is turned on?

   There is an open in the bulb circuit that causes the instrument to see an infinite resistance. The higher the resistance in the bulb circuit, the higher the temperature indication.

8. What type of indicating system is used to measure the exhaust gas temperature of a turbine engine?

   A set of thermocouples arranged in an averaging circuit. These thermocouples are installed in the tail pipe of the engine.

9. What should a thermocouple-type cylinder head temperature indicator read when the engine is not operating?

   The temperature of the outside air.

10. How does a torquemeter actually measure the torque produced by an engine?

    A torquemeter is actually an oil pressure gage. The pressure it measures is produced in a torque sensor and is proportional to the amount of strain in the torsional shaft that drives the reduction gears of the turboprop engine.

11. What two pressures are measured to get the Engine Pressure Ratio of a turbojet engine?

    The turbine discharge total pressure and the compressor inlet total pressure.

12. Of what two materials are the thermocouples made that are used in a turbine engine exhaust gas temperature system?

    Chromel and alumel.

13. What kind of mechanism is normally used to measure oil pressure in a reciprocating engine?

    A bourdon tube mechanism.

14. What is measured by the tachometers used on a two-spool gas turbine engine?

    The N1 tachometer shows the RPM of the low-pressure compressor, and the N2 tachometer shows the RPM of the high-pressure compressor.

15. Does the tachometer of a geared reciprocating engine indicate the speed of the crank shaft or of the propeller shaft?

    The speed of the crankshaft.

Typical Practical Projects

1. Demonstrate to the examiner the correct way to use a Jetcal analyzer to check the thermocouple system used to measure the exhaust gas temperature on a turbine engine.

2. Check the exhaust gas temperature system installed on a reciprocating engine for the condition of the probes and for security of mounting of the thermocouples.

3. Check the fuel flowmeter installed on a fuel-injected horizontally opposed reciprocating engine for the condition of the pressure pickup line and for proper mounting of the instrument.

4. Check a thermocouple-type cylinder head temperature indicator system for the condition of the pickup at the cylinder head, the condition of the lead resistor, and for the correct reading of the indicator when the engine is cold.

5. Check a manifold pressure gage for the correct reading of the instrument when the engine is not running.

6. Check the markings on the engine instruments installed in an aircraft to determine whether or not these markings agree with the engine limitations specified in the appropriate Type Certificate Data Sheets.

7. Inspect the shock mounts that support the instrument panel in an aircraft for their condition.

8. Explain to the examiner the way an EGT indicator can assist the pilot in getting the proper fuel-air mixture ratio for engine operation.

9. Check the tachometer cable used on a light aircraft reciprocating engine for security of mounting and for evidence of kinking.

10. Explain to the examiner what type of electrical power is needed to operate Magnesyn and Autosyn remote indicating instrument systems.

11. Explain to the examiner the correct meaning of each of the colored limitation marks on an aircraft engine instrument.

ENGINE FIRE PROTECTION SYSTEMS

Study Materials

Aviation Maintenance Technician Series
Powerplant textbook Pages 609–628
ASA, Inc.

Airframe and Powerplant Mechanics Powerplant Handbook AC 65-12A Pages 391–409
Federal Aviation Administration

Typical Oral Questions

1. Can a thermal switch-type fire detection system indicate a general overheat condition?

   No, this is a spot-type fire detection system, and it cannot indicate an overheat condition.

2. Can a thermocouple-type fire detection system indicate a general overheat condition?

   No, this is a rate-of-temperature-rise-type of system, and it cannot indicate a general overheat condition.

3. How is a thermocouple-type fire detection system checked for operation?

   Current is sent through the heater in the thermal test unit. This heats the thermocouple, which produces enough current to close the sensitive relay. Current through the contacts of the sensitive relay closes the slave relay and turns on the fire warning light.

4. Does a break in a continuous-loop fire detection system keep the system from detecting a fire?

   No, the unit will show a fault when it is tested, but it will still warn of a fire.

5. Does the pressure-type continuous-element fire detector system detect a general overheat condition, as well as a fire?

   Yes, a high temperature on a small portion of the sensor or a lower temperature over the entire sensor will cause it to release enough gas to close the diaphragm switch and warn of a fire.

6. What are the two most commonly used fire extinguishing agents used for engine fires?

   Halogenated hydrocarbon, such as Halon 1301, and carbon dioxide.

7. What extinguishing agent is used in the high-rate-discharge fire extinguishing systems installed in most turbojet engine installations?

   Halon 1301.

8. What happens when the pilot pulls the fire-pull T-handle in the cockpit of a jet transport airplane?

   The bottle discharge switch is uncovered and armed, the generator field relay is tripped, fuel is shut off to the engine, hydraulic fluid is shut off to the pump, the engine bleed air is shut off, and the hydraulic pump low-pressure lights are deactivated.

9. How can a mechanic determine the state of charge of the HRD bottles of fire extinguishing agent?

   A pressure gage on the bottle shows the pressure of the contents of the bottle.

10. What method is used to discharge an HRD bottle in a turbine engine installation?

    An electrically ignited powder charge blows a knife through the seal on the bottle and discharges the contents.

11. Which type of fire detection system operates on the rate of temperature rise?

    The thermocouple system.

12. What is indicated if the yellow blow-out plug of the fire extinguisher system on the side of an engine nacelle is blown out?

    The built-in fire extinguishing system has been discharged by normal operation.

13. What is indicated if the red blow-out plug of the fire extinguisher system on the side of an engine nacelle is blown out?

    The built-in fire extinguishing system has been discharged by a high temperature condition.

14. How is a pressure-type continuous-element fire detector system checked for operation?

    Low-voltage alternating current is sent through the sheath of the detector element. This heats the element, and it releases enough gas to close the diaphragm switch.

Typical Practical Projects

1. Locate and identify the fire protection system components in an aircraft specified by the examiner.

2. Examine the HRD bottles installed in an aircraft and determine whether or not they have sufficient charge in them.

3. Examine a portable CO_2 fire extinguisher to determine whether or not it contains a full charge, and when it was last serviced.

4. Locate and identify to the examiner the blow-out plugs on the side of an engine nacelle, and explain the meaning of each of them.

5. Check a fire detection system specified by the examiner for operation. Explain to the examiner what is taking place when the system is being checked.

6. Examine a continuous-loop fire detection system, and explain to the examiner the types of damage that are likely to be found in this system.

7. Locate and identify the explosive cartridge that is used to discharge an HRD fire extinguishing agent bottle. Explain to the examiner the way to determine the service life of the cartridges.

8. Demonstrate to the examiner the correct way to use a hand-held fire extinguisher to extinguish an induction system fire in a reciprocating-engine-powered airplane.

ENGINE ELECTRICAL SYSTEMS

Study Materials

Aviation Maintenance Technician Series
Powerplant textbook — Pages 581–608
ASA, Inc.

Airframe and Powerplant Mechanics Powerplant Handbook AC 65-12A — Pages 235–262
Federal Aviation Administration

Advisory Circular 43.13-1B — Pages 11-1–11-118
Federal Aviation Administration

Typical Oral Questions

1. Why is stranded wire used rather than solid wire in most powerplant electrical systems?

   Solid wire is likely to break when it is subjected to vibration.

2. What two things must be considered when selecting the size wire to use when making an electrical installation in an aircraft?

   The current-carrying capability of the wire and the amount of voltage drop that is caused by current flowing through the wire.

3. Why are the wires in certain electrical installations twisted together?

   By twisting the wires together, the magnetic fields caused by current flowing in the wires are minimized.

4. What is used to protect a wire bundle from chafing where it passes through a hole in a bulkhead or frame?

   A grommet around the edges of the hole.

5. How are electrical wires protected where they pass through an area of high temperature?

   Wires passing through these areas are insulated with high temperature insulation, and the wires are enclosed in some type of protective conduit.

6. What is the minimum separation that is allowed between a wire bundle and a fluid line that carries combustible fluid or oxygen?

   A minimum of six inches.

7. What is the significance of the color of the solderless connectors that are used on electrical wires?

   The color of the insulation indicates the size of wire the connector will fit. Red terminals fit 22- through 18-gage wire, blue terminals fit 16- and 14-gage wire, and yellow terminals fit 12- and 10-gage wire.

8. What is the maximum number of wire terminals that may be stacked on a single stud in a terminal strip?

   Four.

9. What is the purpose of the shielding that is used to encase some electrical wires?

   Shielding intercepts radiated electromagnet energy and carries it to ground so it will not interfere with any nearby sensitive electronic equipment.

10. Why must a switch be derated if it is used in a circuit controlling a DC electric motor?

    The initial current flowing into a DC electric motor is much higher than the current the motor uses after the armature begins to rotate. Because of this high inrush current, the controlling switches must be derated.

11. What is used as the rectifier to produce direct current in a DC generator?

    Brushes and a commutator.

12. How does a vibrator-type voltage regulator control the output voltage of a DC generator?

    The strength of the magnetic field produced in the voltage regulator relay is proportional to the generator output voltage. When the voltage rises above the regulated value, the relay pulls the contacts open and inserts a resistor in the generator field circuit. The contacts vibrate open and closed, putting the resistor in and out of the field circuit to control the amount of voltage the generator produces.

13. What is meant by paralleling the generators of a twin-engine aircraft?

    Adjusting the voltage of the generators so they will share the electrical load equally.

14. What is meant by flashing the field of a DC generator?

    Restoring the residual magnetism to the frame of a generator by passing battery current through the field coils in the same direction it flows when the generator is producing current.

15. What kind of rectifier is used in a DC alternator of the type that is used on most of the modern light airplanes?

    A full-wave, three-phase rectifier made up of six silicon diodes.

16. What is used to maintain a constant frequency of the alternating current that is produced by an AC alternator driven by an aircraft turbine engine?

    A hydraulic constant-speed drive unit between the engine and the alternator.

17. What three things must be synchronized before a three-phase AC generator can be connected to a bus that is being served by another generator?

    The voltage, the frequency, and the phase rotation of the generators.

18. What must be done to reverse the direction of rotation of the armature of a DC electric motor?

    The current flow must be reversed through the armature or the field windings, but not through both of them.

19. Does a series-wound DC motor have a high or a low starting torque?

    A high starting torque.

20. Which aircraft electrical circuit does not normally contain a fuse or circuit breaker?

    The starter motor circuit.

21. When removing a battery from an aircraft, which connection should be removed first?

    The ground connection must be disconnected first and connected last.

22. What is meant by the effective voltage of sine wave alternating current?

    The value of the alternating current that is needed to produce the same amount of heat as this value of direct current.

23. What is a starter-generator that is used with many of the smaller gas turbine engines?

    A single-engine-mounted component that serves as a starter for starting the turbine engine. When the engine is running, the circuitry can be shifted so it acts as a compound-wound generator.

24. When should aircraft wiring be installed in a conduit?

    When the wiring passes through an area in the aircraft where open wiring could likely be damaged, such as through a wheel well.

Typical Practical Projects

1. Disassemble, clean, and check an aircraft generator according to the service instructions furnished by the generator manufacturer.

2. Check the armature of a DC generator on a growler. Turn the commutator, and undercut the mica according to the instructions furnished by the generator manufacturer.

3. Install and properly seat the brushes in a DC generator. Check the brush tension to determine whether or not it is within the tolerance allowed by the generator manufacturer.

4. Explain to the examiner the correct way to flash the field of a DC generator.

5. Use the illustrated parts catalog for an aircraft to find the parts numbers of a series of components specified by the examiner.

6. Explain to the examiner the correct way to parallel the generators that are installed on a twin-engine aircraft.

7. Select the correct solderless terminal for an electrical wire and install it, using the proper tool.

8. Check the diodes in a DC alternator for shorts or opens.

9. Using a multimeter, check the continuity in an electrical circuit specified by the examiner.

10. Install wires in a quick-disconnect connector. Use the proper insulation, and secure the wires with the proper wire clamp.

11. Using the wiring diagram of an aircraft generator system, explain to the examiner the probable causes of the generator failing to keep the aircraft battery fully charged.

12. Using the aircraft illustrated parts catalog, find the correct part number of a starter solenoid. Explain to the examiner the difference between a starter solenoid and a battery master solenoid.

13. Using the wire chart in AC 43.13-1B, find the wire size needed to carry a specified amount of current for a specified distance without exceeding the allowable current-carrying limits or the allowable voltage drop.

LUBRICATION SYSTEMS

Study Materials

Aviation Maintenance Technician Series
Powerplant textbook   Pages 87–117 and 417–442
ASA, Inc.

Airframe and Powerplant Mechanics Powerplant Handbook AC 65-12A   Pages 285–314
Federal Aviation Administration

Typical Oral Questions

1. What is meant by the viscosity of engine lubricating oil?

   The resistance of the oil to flow.

2. What is meant by a wet sump lubrication system?

   A lubrication system in which the oil is carried inside the engine itself.

3. What is meant by a dry sump lubrication system?

   A lubrication system in which the oil is carried in a tank that is not a part of the engine.

4. How is oil temperature controlled in an aircraft reciprocating engine?

   Hot oil is directed through the core of the oil cooler, but cold oil is directed around the outside of the core so it will not be further cooled.

5. Is the oil temperature shown on the aircraft instrument panel the temperature of the oil entering the engine or of the oil leaving the engine?

   It is the temperature of the oil entering the engine.

6. What is the purpose of oil dilution in a reciprocating engine?

   When very cold weather is anticipated, gasoline can be mixed with the lubricating oil before the engine is shut down. This reduces the viscosity of the oil and makes starting easier. When the engine is running, the gasoline evaporates out of the oil.

7. What is the function of the hopper in the oil tanks used with some reciprocating engines?

   Hoppers are a part of the oil dilution system. Only the oil in the hopper is diluted. This speeds up dilution and requires less gasoline.

8. What happens to the gasoline that is used to dilute the oil in the crankcase of an aircraft reciprocating engine?

   When the engine is running and the oil is warm, the gasoline evaporates out of it.

9. What is the purpose of the sludge plugs in the crankshaft of a reciprocating engine?

   They trap sludge that is in the oil and hold it until the engine is disassembled at overhaul.

10. Which pump is the larger in a dry-sump lubricating system, the pressure pump or the scavenger pump?

    The scavenger pump has the greater volume.

11. Why do full-flow oil filters have a spring-operated bypass valve in them?

    In case the filter should plug up so it cannot pass any oil, the bypass valve will open and allow unfiltered oil to flow through the system.

12. What is the function of a fuel-oil heat exchanger in the lubrication system of a turbojet engine?

    This allows heat from the oil to warm the fuel so ice will not form on the fuel filters.

13. What kind of oil is used in most turbojet engines?

    Synthetic oil.

14. Where are the last chance oil filters located in a turbojet engine?

    They are located inside the engine just ahead of the nozzles that spray oil onto the bearings.

15. Where is the oil tank in a dry-sump reciprocating engine vented?

    To the engine crankcase.

16. What information must be displayed around the oil filler opening for a turbojet engine?

    The word "Oil" and the permissible oil designations, or references to the Airplane Flight Manual (AFM) for permissible oil designations.

17. What is the function of the oil control rings on the piston of an aircraft reciprocating engine?

    They maintain the proper quantity of oil between the piston and the cylinder wall.

18. What is meant by a spectrometric oil analysis?

    A type of oil analysis in which a sample of oil is burned in an electric arc. The wavelengths of the light produced by the burning oil are analyzed, and this type of analysis shows the number of parts per million of the various metals that are in the lubricating oil.

19. What indication would a pilot have if his oil supply was low?

    The oil temperature would be high and the oil pressure would be low.

20. What is meant by a hot-tank lubrication system?

    A lubrication system in which the oil cooler is located in the pressure subsystem.

21. What is meant by a cold-tank lubrication system?

    A lubrication system in which the oil cooler is located in the scavenge subsystem. The oil that is returned to the tank has been cooled.

Typical Practical Projects

1. Demonstrate the correct way to adjust the oil pressure on an engine specified by the examiner.

2. Demonstrate the correct way to preoil an engine specified by the examiner.

3. Using a diagram of a turbine engine lubrication system, identify the filters, the spray nozzles, the pumps, the relief valves, the check valves, and the bypass valves.

4. Using the Type Certificate Data Sheets for an aircraft specified by the examiner, find the oil quantity specified and the amount of undrainable oil that is trapped in the system.

5. Remove an oil strainer from an engine, clean and inspect it, and reinstall it in the engine.

6. Demonstrate to the examiner the proper way to change the oil filter on an aircraft engine. Explain the precautions that must be taken to keep from damaging the engine when installing the new filter.

7. Demonstrate to the examiner the correct way to inspect an oil filter that has been removed from an aircraft engine.

8. Demonstrate to the examiner the correct way to inspect an engine for the source of an oil leak.

9. Using a list of oils that are approved for an aircraft engine, choose the correct oil for the existing climatic conditions and explain to the examiner the reason for your choice.

10. Demonstrate to the examiner the correct way to dilute the oil in an aircraft reciprocating engine prior to shutting the engine down.

11. Explain to the examiner the way a Cuno filter operates.

IGNITION SYSTEMS

Study Materials

Aviation Maintenance Technician Series Powerplant textbook Pages 193–238 and 479–492
ASA, Inc.

Airframe and Powerplant Mechanics Powerplant Handbook AC 65-12A Pages 177–235
Federal Aviation Administration

Typical Oral Questions

1. What is the main advantage of a magneto ignition system over a battery ignition system for an aircraft reciprocating engine?

   A magneto has its own source of electrical energy, and it is not dependent upon the battery.

2. What is the function of the capacitor in a magneto?

   The capacitor minimizes arcing at the breaker points, and it speeds up the collapse of the primary current as the breaker points open.

3. What is a compensated magneto cam, and on what kind of engine is one used?

   A compensated cam is a special cam used in magnetos mounted on high-performance radial engines. The cam has one lobe for each cylinder, and the lobes are ground in such a way that the breaker points open when the pistons in the different cylinders are the same linear distance from top center.

4. What is the significance of the numbers on the distributor of an aircraft magneto?

   These numbers are the sparking order of the magneto, not the firing order of the engine.

5. What is done in a magneto ignition system when the ignition switch is placed in the Off position?

   The primary circuit is connected to ground.

6. What is the reason for having a low-tension ignition system on some aircraft?

   Low-tension magnetos are used on aircraft that fly at high altitudes where there is a problem with flashover in the high-tension magneto distributor.

7. What is the basic difference between a low-tension magneto and a high-tension magneto?

   The low-tension ignition system has only one coil in the magneto, and it uses a carbon-brush-type distributor. Low voltage is distributed to high-tension transformers that are located on the heads of each cylinder.

   A high-tension magneto produces high voltage in the magneto coil, and it is sent to the correct spark plug by the built-in high-voltage distributor.

8. What is meant by an All Weather spark plug?

   A shielded spark plug that has a recess in the shielding in which a resilient grommet on the ignition lead forms a watertight seal.

9. What is meant by the reach of a spark plug?

   The length of the threads on the spark plug that screw into the cylinder head.

10. What is the difference between a hot spark plug and a cold spark plug?

    A hot spark plug has a long path for the heat to travel between the nose core insulator and the spark plug shell. In a cold spark plug, the heat has a shorter distance to travel, and the spark plug operates cooler than a hot spark plug.

11. What is checked when a magneto is internally timed?

    Internally timing a magneto consists of adjusting the breaker points so they will open at the instant the rotating magnet is in its E-gap position, and the distributor rotor is in the position to direct the high voltage to cylinder number one.

12. What is the purpose of a vernier coupling used on some aircraft magneto drives?

    Magnetos that are base mounted must have a vernier coupling between the magneto drive and the engine. This vernier coupling allows the magneto-to-engine timing to be varied in increments of less than one degree.

13. What is the advantage of fine-wire spark plugs over massive electrode spark plugs?

    Fine-wire spark plugs have a firing end that is more open than that of a massive electrode spark plug. The open firing end allows the gases that contain lead to be purged from the spark plug so they will not form solid lead contaminates.

14. Why is it important that the spark plugs be kept in numbered holes in a tray when they are removed from an engine?

Spark plugs tell a good deal about the internal condition of the cylinders from which they were taken. By knowing the cylinder from which each spark plug came, the mechanic can take the proper action when a spark plug indicates such conditions as detonation or overheating.

15. What is the purpose of staggered timing between the two magnetos on an aircraft engine?

Engines in which the exhaust gas scavenging from the cylinders is uneven use staggered ignition timing. The spark plug nearest the exhaust valve, where the fuel-air mixture is diluted, fires before the spark plug on the intake side. By using staggered timing, the flame front caused by the two spark plugs will meet in the center of the piston.

16. In what position should the ignition switch be placed when using a timing light on the magnetos?

In the Both position.

17. Why is it important that a torque wrench always be used when installing spark plugs in an aircraft engine?

If the spark plugs are not put in tight enough, there is the possibility of a poor seal, and if they are put in too tight, there is danger of cracking the insulation.

18. Why are pressurized magnetos used on most reciprocating-engine-powered aircraft that fly at high altitudes?

Pressurized air is a better insulator than less dense air. By pressurizing the distributors, the high voltage is kept from arcing across to the wrong electrode and causing vibration and loss of engine power.

19. What type of ignition system is used on most turbine engines?

High-intensity, intermittent-duty, capacitor discharge ignition systems.

20. What is meant by the E-gap in magneto timing?

The E-gap angle is the position of the rotating magnet when the primary current flowing in the magneto coil is the greatest. The breaker points open when the rotating magnet is in its E-gap position.

21. What is the function of an impulse coupling?

An impulse coupling is a spring-driven coupling between the magneto and the engine. When the engine is being started, the impulse coupling holds the rotating magnet until the piston passes over its top center position and starts down. The impulse coupling releases the magnet and the spring spins it fast so that it produces a hot and late spark.

22. What kind of gage should be used to measure the electrode gap in aircraft spark plugs?

A round wire gage.

23. How many igniters are used in most turbine engines?

Two.

24. What are two types of ignition systems used in turbine engines?

High-voltage systems and low-voltage systems.

25. With which type of ignition system is a glow plug igniter used?

A low-voltage system.

26. How is the strength of the magnet in a magneto checked?

The magneto is put on a test stand and rotated at a specified speed. The breaker points are held open and the primary current is measured. The strength of the magnet determines the amount of primary current.

27. In what position is the magnet in a magneto when the greatest change in flux density in the coil core takes place?

It is a few degrees beyond its neutral position. When it is in this position, the breaker points open and the primary current is interrupted. The flux change in the coil core is the greatest.

28. In what position is the magnet in a magneto when the breaker points begin to open?

In its E-gap position, just a few degrees beyond its neutral position.

29. What malfunction in the ignition system would cause an aircraft reciprocating engine to continue to run after the ignition switch is placed in the Off position?

The ignition switch is not grounding the magneto primary circuit.

30. What turns on the autoignition system in a turboprop engine?

A torque pressure switch energizes the system when the engine stops producing torque.

Typical Practical Projects

1. Check the strength of the magnet in an aircraft magneto and determine whether or not it meets the manufacturer's specifications.

2. Demonstrate to the examiner the correct way to internally time an aircraft magneto.

3. Using the proper information, place the magnet of a magneto in its E-gap position.

4. Demonstrate to the examiner the correct way to time a magneto to an aircraft engine.

5. On an engine specified by the examiner, perform an engine run-up that includes a check of the magnetos and an ignition system safety check.

6. Check a high-tension magneto lead for its physical and its electrical condition.

7. Draw a schematic of the ignition of an aircraft magneto ignition system. Include the coil, the condenser, the breaker points, the distributor, the spark plugs, and the ignition switch.

8. Using the correct test equipment, check the condition of the condenser in an aircraft magneto.

9. Demonstrate to the examiner the correct way to clean, gap, and test aircraft spark plugs.

10. Using the correct publications, compile a list of spark plugs that are approved for an aircraft engine specified by the examiner.

11. Demonstrate to the examiner the correct way to inspect and clean a turbine engine igniter.

12. Using the proper reference material, explain to the examiner the correct way to inspect the ignition system of a turbine engine.

13. Using the proper publications, compile a list of igniters that are approved for an aircraft turbine engine specified by the examiner.

FUEL METERING SYSTEMS

Study Materials

Aviation Maintenance Technician Series Powerplant textbook Pages 119–177 and 447–478
ASA, Inc.

Airframe and Powerplant Mechanics Powerplant Handbook AC 65-12A Pages 109–175
Federal Aviation Administration

Typical Oral Questions

1. Where does the fuel metering system of a turbine engine discharge its fuel?

   The fuel is discharged through spray nozzles into the combustion chambers.

2. What is adjusted when the fuel control unit of a turbine engine is trimmed?

   The idle speed and the maximum-thrust speed.

3. What are two locations water may be injected into a turbine engine?

   At the compressor inlet and at the inlet to the diffuser section.

4. What is meant by a duplex nozzle in a turbine engine fuel metering system?

   A duplex fuel nozzle is one that has two fuel discharge passages. A flow divider sends fuel for low-pressure operation through one discharge passage, and when the engine demands more fuel, it is sprayed out through the second discharge passage. The spray pattern keeps the flame centered in the burner for all operating conditions.

5. Does the fuel-air mixture provided by a float carburetor become richer or leaner as the aircraft goes up in altitude?

   If the mixture is not adjusted, it will become richer as the aircraft gains in altitude.

6. Does the fuel-air mixture provided by a float carburetor become richer or leaner when carburetor heat is applied?

   The mixture becomes richer when carburetor heat is applied.

7. What would happen to the fuel-air mixture ratio in a float carburetor if the main air bleed were to become plugged?

   The mixture would become excessively rich.

8. How does the automatic mixture control in a pressure carburetor keep the fuel-air mixture constant as the aircraft changes altitude?

   It automatically, and progressively, bleeds the air between the two sides of the air diaphragm and decreases the air metering force. This leans the mixture as the aircraft goes up in altitude.

9. What are two purposes of the manifold valve (or flow divider) in the fuel injection system of a horizontally opposed aircraft engine?

   It provides a constant discharge fuel pressure for idling, and it provides a positive shutoff for the fuel when the engine is shut down.

10. How does the mixture control change the fuel-air mixture ratio on the Teledyne-Continental fuel injection system?

    When the mixture control is in its Full Rich position, all fuel goes to the manifold valve. When it is in the Idle Cutoff position, all fuel is returned to the pump inlet. Intermediate positions vary the amount of fuel that goes to the engine.

11. Why must the diaphragms of a pressure carburetor be soaked before the carburetor is ready for flight?

    Soaking the diaphragms of a pressure carburetor restores them to the condition of flexibility that they had when the carburetor was calibrated.

12. What are two basic types of fuel controls for aircraft turbine engines?

    Hydropneumatic and electro-hydromechanical.

13. How does the exhaust gas temperature of a reciprocating engine tell anything about the fuel-air mixture being burned by the engine?

    A stoichiometric mixture (a mixture in which all of the constituents of the fuel-air mixture are burned) produces the highest exhaust gas temperature. The mixture is adjusted to get the highest EGT, and then it is enriched to place it on the rich side of stoichiometric.

14. What two things are adjusted when adjusting the idling of a reciprocating engine?

    The throttle stop, to get the proper idling RPM, and the idle mixture control, to get the smoothest operation.

15. What is controlled by the adjustment of the orifice in the fuel pump of a Teledyne-Continental fuel injection system?

    The high unmetered fuel pressure.

16. What is the significance of the letters that are stamped on the flats of a fuel injection nozzle?

    These letters designate the relative size of the orifice in the injector nozzle. The lower the letter in the alphabet, the smaller the amount of fuel the nozzle will flow.

17. What is the function of the derichment valve in a pressure carburetor used on an aircraft engine with an antidetonation injection system?

    The derichment valve automatically closes to lean the fuel-air mixture when ADI fluid is flowing. As soon as the ADI fluid stops flowing, the derichment valve opens and the mixture returns to its rich condition.

18. What is the function of the economizer system in an aircraft engine carburetor?

    The economizer (or power enrichment) system allows the engine to operate with an economically lean mixture for all conditions other than full power. When the throttle is opened for full power, the power enrichment system automatically enriches the mixture to remove some of the heat that is developed during this type of operation.

19. In what position is the mixture control placed for starting an engine that is equipped with a pressure carburetor?

    In the Idle Cutoff position. The engine is started with fuel from the primer system.

20. In what position is the carburetor heat control placed when starting an aircraft engine?

    In the Cold position.

21. Why is an aircraft reciprocating engine killed by placing the mixture control in the Cutoff position, rather than by using the magneto switch?

    By shutting off the fuel to the cylinders, the combustion chambers are left full of air with no fuel. This makes the engine less likely to fire if the propeller is pulled through with the ignition switch accidentally on.

22. Where is the fuel from a continuous-flow fuel injection system discharged?

    It is discharged through injector nozzles screwed into the cylinder heads near the intake valves.

23. In which direction relative to the wind should a turbine-powered aircraft be positioned when the fuel control is being trimmed?

    If the wind velocity is less than 10 miles per hour, it can be faced in any direction. If the wind velocity is between 10 and 25 MPH, it should be trimmed facing into the wind. The engine should not be trimmed when the wind velocity is more than 25 MPH.

24. What engine parameters are sensed by the fuel control unit of a turbine engine?

    Engine RPM, inlet air pressure, compressor discharge pressure, burner can pressure, and inlet air temperature.

25. When does the acceleration system operate on an aircraft carburetor?

    When the throttle is suddenly opened.

Typical Practical Projects

1. Using the correct reference material, adjust the float level in an aircraft carburetor.
2. Install a float carburetor on an aircraft engine specified by the examiner. Adjust the idling RPM and mixture, and check to determine that the engine develops the proper static RPM.
3. Using the correct reference material, determine whether the metering jet and the venturi in a carburetor specified by the examiner are correct.
4. Demonstrate to the examiner the way to determine whether or not the idling RPM and mixture are correct on a fuel-injected reciprocating engine.
5. Remove and clean the fuel strainer in a carburetor or fuel injection system specified by the examiner.
6. Remove and properly clean a set of injector nozzles from an engine specified by the examiner. Explain to the examiner the points to check when cleaning and installing the nozzles.
7. Using a diagram of a pressure carburetor, explain to the examiner the way the air metering and the fuel metering forces are produced.
8. Using a diagram of a Bendix fuel injection system, explain to the examiner the way the air metering and the fuel metering forces are produced.
9. Using a diagram of a Teledyne-Continental fuel injection pump, explain to the examiner the way the orifice controls the high unmetered fuel pressure.
10. Locate and identify on a turbine engine fuel control, the controls that are adjusted to trim the engine.
11. Explain to the examiner the correct way to soak a pressure carburetor after it is installed on an engine.
12. Explain to the examiner the way the idling RPM and the idle mixture ratio is adjusted on a pressure carburetor or fuel injection system specified by the examiner.
13. Using a diagram of a water injection system installed on a turbine engine, explain to the examiner the way the system operates.

ENGINE FUEL SYSTEMS

Study Materials

Aviation Maintenance Technician Series
Airframe textbook, Volume 2 Pages 587–649
ASA, Inc.

Title 14 of the Code of Federal Regulations, Part 23
Federal Aviation Administration

Typical Oral Questions

1. What is the purpose of the bypass valve in an engine-driven fuel pump?

   The bypass valve allows fuel to flow around the engine-driven pump for starting and for emergency operation if the engine-driven pump should fail.

2. On which side of the firewall is the engine fuel shutoff valve located?

   The shutoff valve must be on the side of the firewall away from the engine.

3. What are three purposes for the boost pumps in an aircraft fuel system?

   To provide fuel pressure for starting the engine, to pressurize the fuel lines to prevent vapor lock, and to transfer fuel from one tank to another.

4. What kind of boost pump is used in most aircraft fuel systems?

   Electrically operated centrifugal pumps.

5. Where does an engine-driven fuel pump direct the excess fuel from its pressure relief valve?

   Back to the inlet side of the pump.

6. What causes a vapor lock in an aircraft fuel system?

   The fuel becomes hot enough that it boils. Vapors are released from the liquid fuel, and these vapors block the fuel lines so the liquid fuel cannot flow to the engine.

7. What is done in most aircraft fuel systems to prevent vapor lock?

   Boost pumps in the fuel tank pressurize the fuel in the lines and force the fuel into the fuel metering system.

8. What is meant by a compensated relief valve in an engine-driven fuel pump?

   A pressure relief valve that is acted upon by a diaphragm as well as a spring. Atmospheric pressure acting on the diaphragm varies the pump discharge pressure so that it will remain a given amount higher than the pressure of the air entering the carburetor.

9. What is the purpose of the pressurizing and dump valve in the fuel system for a turbine engine?

   For normal engine operation, the pressurizing and dump valve acts as a flow divider, directing the fuel into the main or pilot manifold so it will be discharged from the proper orifice in the duplex fuel nozzle.

   When the engine is shut down, the dump function of the valve dumps all of the fuel from the manifold.

10. What should be done to an aircraft reciprocating engine if it has been operated on turbine engine fuel?

    All of the turbine fuel should be drained out and the system filled with the proper grade of aviation gasoline. The engine should be given a compression check, and all of the cylinders should be inspected with a borescope. The oil should be drained and the filters carefully examined. The engine should be given a complete run-up check.

11. What are the two basic types of turbine engine fuel?

    Jet A and A-1, which are a special type of kerosine-base fuel. This is similar to military JP-5.

    Jet B is a gasoline-base fuel similar to military JP-4.

12. Why do some turbine engine fuel systems incorporate a fuel heater?

    The fuel heater keeps the fuel warm enough that any water that precipitates out of the fuel will not freeze on the filters.

13. What are two reasons Prist is used in turbine engine fuel?

    Prist is an antifreeze agent that lowers the freezing point of any water that precipitates out of the fuel. It also acts as a biocidal agent that kills the microbial growth that forms scum in fuel tanks. This scum traps and holds water against the aluminum alloy in the fuel tanks and causes corrosion.

14. Why do some aircraft fuel filters have a built-in relief valve?

    This relief valve will open and allow unfiltered fuel to flow to the fuel control device if the filter should become plugged with ice or other contaminants.

Typical Practical Projects

1. Remove and clean a fuel strainer, reinstall it, and pressure check it for leaks.

2. Explain to the examiner the correct way to adjust the pressure on an engine-driven fuel pump.

3. Using a diagram of an engine-driven fuel pump, explain to the examiner the operation of the bypass valve and the pressure relief valve.

4. Using the proper reference material, explain to the examiner the way a submerged boost pump should be removed from an aircraft fuel tank.

5. Inspect a fuel selector valve in an aircraft, and determine if each position of the valve can be positively identified by feel.

6. Locate and identify the fuel strainers in the fuel system of a turbine engine.

7. Locate and identify the fuel heater in the fuel system of a turbine engine.

ENGINE INDUCTION SYSTEMS

Study Materials

Aviation Maintenance Technician Series Powerplant textbook — Pages 178–190 and 376–379
ASA, Inc.

Airframe and Powerplant Mechanics Powerplant Handbook AC 65-12A — Pages 71–95
Federal Aviation Administration

Typical Oral Questions

1. Where is the heat taken from that is used to heat the induction air in a reciprocating engine?

   From a muff that is installed around some part of the exhaust system

2. Where does the alternate air come from that is used with a pressure carburetor or a fuel injection system?

   From inside the engine cowling.

3. Where does carburetor ice normally form in a carburetor?

   In the throat of the carburetor, on and around the throttle valve.

4. Does the application of carburetor heat cause the fuel-air mixture to become richer or to become leaner?

   Heated air causes the fuel-air mixture to become richer.

5. What is used to drive most of the external superchargers used on modern reciprocating engines?

   Exhaust gases.

6. What controls the speed of a turbocharger compressor?

   The amount of exhaust gas that is forced to flow through the turbine. This is controlled by the position of the waste gate.

7. What is meant by a convergent inlet duct for a turbine engine?

   A convergent duct is one whose cross-sectional area becomes smaller in the direction the air flows.

8. What is meant by a divergent inlet duct for a turbine engine?

   A divergent duct is one whose cross-sectional area becomes greater in the direction the air flows.

9. What kind of inlet duct is often used on turbine-powered helicopters?

   A bell-mouthed inlet duct.

10. What is the danger of operating an aircraft reciprocating engine with too high a carburetor air temperature?

    Too high a carburetor air inlet temperature can cause the fuel-air mixture to reach its critical temperature and detonate.

11. Why do some turbine engines use variable inlet guide vanes?

    These variable inlet guide vanes are automatically adjusted to direct the air into the engine in such a way that it keeps the RPM vs. velocity proper for the most efficient operation.

12. How do some turbine engines prevent ice formation on the inlet guide vanes?

    Hot compressor bleed air flows through hollow inlet guide vanes.

13. What is usually installed in a large reciprocating engine between the turbosupercharger and the carburetor?

    An intercooler. This is an air-to-air heat exchanger.

14. In what position should the carburetor heat control be placed when starting a reciprocating engine?

    In the Cold position.

15. How does an engine air inlet vortex destroyer help prevent foreign object damage to the engine?

    A high-velocity stream of compressor bleed air is blown out in front of the engine to break up the vortices that form in front of the engine when it is operating at high power on the ground.

Typical Practical Projects

1. Demonstrate to the examiner the correct way to clean the carburetor air filter for an aircraft engine.

2. Inspect a carburetor heater air box, and explain to the examiner the things that should be checked.

3. Demonstrate to the examiner the things that should be inspected on a turbocharger installation.

4. Inspect the induction system of a reciprocating engine for condition. Explain to the examiner the things that could cause a loss of engine power.

5. Locate and identify the components in the control system of a turbocharger.

6. Using a diagram of the ice control system for a turbine engine, explain to the examiner the way each component removes or prevents the formation of ice.

7. Explain to the examiner the correct way to check the operation of a turbocharger system on an aircraft engine.

8. Using a diagram of a turbocharger installation, explain to the examiner the operation of the type of controller that is installed.

9. Using a diagram of the clutch system of a two-speed internal supercharger, explain to the examiner the way the clutches operate to change the speed of the impeller.

10. Explain to the examiner the correct way to use a two-speed supercharger to maintain the maximum engine power as the airplane gains altitude.

11. Explain to the examiner the way the antisurge bleed system in a turbojet engine prevents compressor stall.

12. Explain the way an induction system fire should be extinguished in an aircraft reciprocating engine.

ENGINE COOLING SYSTEMS

Study Materials

Aviation Maintenance Technician Series Powerplant textbook — Pages 253–262 and 443–445
ASA, Inc.

Airframe and Powerplant Mechanics Powerplant Handbook AC 65-12A — Pages 314–324
Federal Aviation Administration

Typical Oral Questions

1. What is meant by pressure cooling of an aircraft engine?

   Air cooling in which air is forced to flow through baffles and cylinder fins by a pressure differential across the engine.

2. What is the purpose of an augmentor tube in the cooling system of an aircraft reciprocating engine?

   The augmentor tube uses the velocity of the exhaust gases to produce a low pressure on one side of the engine that helps pull cooling air through it.

3. Where is the highest temperature located in a turbine engine?

   At the inlet to the high-pressure turbine.

4. How are the turbine inlet guide vanes and the first-stage turbine blades in some turbine engines cooled?

   High-pressure compressor bleed air flows through the hollow guide vanes and hollow turbine blades.

5. Which side of an air-cooled engine cylinder has the greatest amount of cooling fins?

   The side of the cylinder in which the exhaust valve is located.

6. What should be done to repair a bent cooling fin in a cast aluminum cylinder head?

   It is normally best to leave a bent cast fin alone if it does not restrict the flow of air. Cast fins are brittle and could break off.

7. What is used in a helicopter to increase the amount of cooling air that flows over the engine cylinders?

   A belt-driven fan.

8. What should be the position of the cowl flaps on a reciprocating engine when it is run on the ground?

   They should be wide open when operating the engine on the ground.

9. What is the function of the majority of the air that passes through a turbine engine?

   Most of the air is used for cooling.

10. How does metallic sodium in an exhaust valve aid in transferring heat?

    The sodium melts when the engine is running, and it sloshes up and down as the valve operates. The sodium picks up heat from the valve head and carries it into the stem so it can be transferred to the cylinder head and dissipated into the air.

11. How is the structure around a turbine engine protected from excessive heat from the engine?

    An insulating blanket protects the structure from excessive heat.

12. What is the function of blast tubes that are installed in a pressure cooling system of a reciprocating engine?

    These blast tubes direct cooling air to the magnetos and the generator.

13. Of what material is the insulation blanket made that is used to protect the aircraft structure from the heat produced by a turbine engine?

    Fiberglass sandwiched between sheets of metal foil.

Typical Practical Projects

1. Examine the baffles in a reciprocating engine installation. Explain to the examiner the things in the baffle system that could cause engine damage.

2. Using the proper reference materials, repair damaged fins on an air-cooled engine cylinder.

3. Inspect the cowl flap system of an air-cooled engine and determine whether or not the flaps reach their fully open and fully closed positions.

4. Using a drawing of the augmentor system installed on a reciprocating engine, explain to the examiner the way this system aids in the cooling of the engine.

5. Explain to the examiner the correct way to inspect the cooling fan and baffles installed with a helicopter engine.

6. Explain to the examiner the correct way to inspect the insulating blanket around a turbine engine.

7. Using a drawing of the cooling system of a turbine engine, explain to the examiner the way cooling air removes heat from the engine.

8. Inspect an air-cooled engine cylinder, and explain to the examiner the things that would indicate that the cylinder has been improperly cooled.

ENGINE EXHAUST SYSTEMS

Study Materials

Aviation Maintenance Technician Series Powerplant textbook Pages 239–252 and 501– 512
ASA, Inc.

Airframe and Powerplant Mechanics Powerplant Handbook AC 65-12A Pages 80–107
Federal Aviation Administration

Typical Oral Questions

1. Of what material are most reciprocating engine exhaust components made?

   Corrosion-resistant steel.

2. How is the speed of a turbocharger controlled?

   By controlling the amount of exhaust gas that flows through the turbine. This is controlled by the position of the waste gate that is in the exhaust pipe.

3. What type of actuator is used to control the position of the turbocharger waste gate?

   A hydraulic actuator that uses engine oil pressure to move the piston.

4. How does a power recovery turbine increase the power of a reciprocating engine?

   Exhaust-driven velocity turbines are coupled through a hydraulic drive to the engine crankshaft. Energy that would normally be lost is used to drive the turbines, and the turbines assist in turning the crankshaft.

5. How tight should the clamps be tightened that hold sections of a reciprocating engine exhaust system together?

   They should not be so tight that they will be damaged when heat expands the exhaust system components.

6. What would be the effect on engine performance of an internal failure of a muffler?

   This will increase the exhaust back pressure and cause a loss of engine power.

7. What effect on turbine engine operation is caused by a change in the area of the exhaust nozzle?

   Altering the area of the exhaust nozzle affects the compression ratio, the RPM, the mass airflow through the engine, and the EGT.

8. What are two types of thrust reversers that are used with turbojet engines?

   Mechanical blockage (clamshell) and aerodynamic blockage (cascade).

9. How does a noise suppressor reduce the amount of noise produced by a turbojet engine?

   It breaks up the low-frequency vibrations in the exhaust stream and converts them into higher frequencies. These high frequencies are dissipated by the air more easily than low frequencies.

10. What is the purpose of the ball joints and bellows in the exhaust system of a reciprocating engine?

    The ball joints and bellows allow the exhaust system components to change their dimensions without causing any leaks.

11. Why do some supersonic aircraft use a convergent-divergent exhaust duct?

    A CD exhaust duct controls the gas expansion as it leaves the engine so the gases will produce thrust at supersonic velocities.

12. Why is it important that a lead pencil never be used to mark on an aircraft exhaust system?

    The graphite in a lead pencil will infuse into the metal when it gets hot. It makes the metal brittle and likely to crack.

13. What type of damage is normally found in the hot section of a turbine engine?

    Cracks caused by heat.

Typical Practical Projects

1. Using a diagram, explain to the examiner the way an afterburner on a turbine engine operates.
2. Demonstrate to the examiner the correct way to inspect the exhaust system of a reciprocating engine for leaks.
3. Locate and identify the components in the exhaust gas temperature indicating system in a turbine engine.
4. Demonstrate to the examiner the correct way to remove and replace a section of the exhaust system of a reciprocating engine.
5. Demonstrate to the examiner the correct way to clean ceramic-coated components in an aircraft engine exhaust system.
6. Locate and identify the components in a turbocharger system installed on a reciprocating engine.
7. Using a diagram, explain to the examiner the correct way of inspecting the thrust reversers on a turbine engine for proper deployment and proper stowage.
8. Using a diagram, explain to the examiner the way noise suppressors on a turbine engine decrease the noise level.
9. Inspect an engine muffler that is used to supply heat to the aircraft cabin. Explain to the examiner the way the muffler should be checked to be sure that there are no leaks.

PROPELLERS

Study Materials

Aviation Maintenance Technician Series
Powerplant textbook Pages 631–702
ASA, Inc.

Airframe and Powerplant Mechanics Powerplant Handbook AC 65-12A Pages 325–358
Federal Aviation Administration

Typical Oral Questions

1. What type of device is used on propeller blades to remove ice that has formed on them?

   Electrically heated deicer boots.

2. What is used to prevent ice forming on a propeller blade?

   A mixture of ethylene glycol and isopropyl alcohol slung out along the blades.

3. Should an adjustable-pitch propeller be in high pitch or in low pitch for takeoff?

   In low pitch.

4. What is done to cause a Hydromatic propeller to feather?

   High-pressure engine oil is directed into the propeller through the governor.

5. What is done to cause a McCauley propeller to feather?

   Oil is allowed to drain out of the propeller.

6. Does centrifugal twisting moment on a propeller blade tend to move the blades toward high pitch or toward low pitch?

   Toward low pitch.

7. Do the counterweights on a propeller tend to move the blades toward high pitch or toward low pitch?

   Toward high pitch.

8. What is the difference between a controllable propeller and a constant-speed propeller?

   Basically, it is the control system. A controllable-pitch propeller uses a manually operated oil valve to control the pitch, and a constant-speed propeller uses a governor to control the valve.

9. When making a magneto check on an engine equipped with a constant-speed propeller, should the propeller control be in the low-pitch or the high-pitch position?

   It should be in the low-pitch position.

10. What can be done to prevent the front cone bottoming when installing a propeller on a splined shaft?

    Install a spacer behind the rear cone to move the propeller forward on the shaft.

11. What is the purpose of the small holes that are drilled in the end of a wooden propeller?

    These holes vent the inside of the propeller blade and allow moisture that collects in the wood to be released.

12. What is the function of the snap ring inside the hub of a propeller that is mounted on a tapered or splined shaft?

    The snap ring allows the propeller to be pulled off of the shaft when the retaining nut is backed off.

13. What is adjusted inside the governor for a constant-speed propeller to change the speed at which the propeller is operating?

    The compression of the speeder spring.

14. What is the function of the accumulator that is used with some McCauley feathering propellers?

    The accumulator stores oil under pressure when the engine is operating normally. This oil is used to help the propeller blades move toward low pitch when the propeller is being unfeathered.

15. What is meant by the beta range of operation of a turboprop propeller?

    This is the mode of ground operation, and it includes starting, taxiing, and ground reverse operation.

16. What is meant by the alpha range of operation of a turboprop propeller?

    This is the in-flight mode of operation from takeoff to landing.

17. Is the flat surface of a propeller blade the face of the blade or the back of the blade?

    It is the face of the blade.

18. What keeps a McCauley feathering propeller from feathering when the engine is shutdown on the ground?

    A spring-loaded latch mechanism prevents the blades moving into the feather position when the engine is shut down on the ground. In the air, aerodynamic forces keep the propeller rotating, and centrifugal force holds the blades unlatched so they can move to the feather position when oil pressure is taken out of the propeller.

Typical Practical Projects

1. Measure the track of a propeller and, using the correct reference materials, determine whether or not it is within allowable tolerance.
2. Using a universal propeller protractor and the correct reference materials, measure the pitch angle of the blades of a propeller specified by the examiner.
3. Inspect a propeller for damage and explain to the examiner the way to remove small surface damage.
4. Remove a propeller from a splined or tapered shaft. Reinstall it and properly safety it.
5. Explain to the examiner the way front cone bottoming can be detected and corrected.
6. Explain to the examiner the correct way to adjust a propeller governor to get the correct takeoff RPM.
7. Demonstrate to the examiner the correct way to lubricate a propeller. Use the proper reference materials to find the lubricants that are to be used.
8. Using the correct reference material, find out if there is any RPM range that is restricted for an engine-propeller combination specified by the examiner.
9. Inspect a wooden propeller for condition and explain to the examiner the things that could make a wooden propeller unairworthy.
10. Remove and reinstall a propeller on a flanged shaft. Explain to the examiner the precautions that should be taken in this operation.
11. Locate and identify the components in the propeller installation on a turboprop engine.
12. Using a diagram of a propeller governor and the propeller it is used with, explain to the examiner the operation of the governor during on-speed operation, overspeed operation, and underspeed operation.
13. Demonstrate the correct way to check a feathering propeller during ground operation of the engine.